WORKFORCE
DEVELOPMENT
NETWORKS

Dedicated to the memories of
Dr. Anthony (Antonio) Soto,
cofounder of the Center for Employment Training,
and Arthur Pearlroth,
cofounder of the Regional Alliance for Small Contractors

WORKFORCE DEVELOPMENT NETWORKS

Community-Based Organizations
and Regional Alliances

Bennett Harrison
Marcus Weiss

SAGE Publications
International Educational and Professional Publisher
Thousand Oaks   London   New Delhi

For information:

SAGE Publications, Inc.
2455 Teller Road
Thousand Oaks, California 91320
E-mail: order@sagepub.com

SAGE Publications Ltd.
6 Bonhill Street
London EC2A 4PU
United Kingdom

SAGE Publications India Pvt. Ltd.
M-32 Market
Greater Kailash I
New Delhi 110 048 India

Printed in the United States of America

Library of Congress Cataloging-in-Publication Data

Harrison, Bennett.
    Workforce development networks: Community-based organizations and regional alliances / by Bennett Harrison and Marcus Weiss.
        p.   cm.
    Includes bibliographical references and index.
    ISBN 0-7619-0847-1 (acid-free paper). — ISBN 0-7619-0848-X (pbk.: acid-free paper)
        1. Public service employment—United States.   2. Job creation—Government policy—United States.   3. Organizational training—Government policy—United States.   4. Urban youth—Employment—Government policy—United States.   5. Community development—United States.   I. Weiss, Marcus S.   II. Title.
    HD5713.6.U54H367   1998
    331.12'042'0973—dc21   97-33900

This book is printed on acid-free paper.

98 99 00 01 02 03 10 9 8 7 6 5 4 3 2 1

Acquiring Editor:	Catherine Rossbach
Editorial Assistant:	Kathleen Derby
Production Editor:	Sherrise M. Purdum
Production Assistant:	Denise Santoyo
Typesetter/Designer:	Janelle LeMaster
Indexer:	Teri Greenberg
Cover Designer:	Candice Harmon
Print Buyer:	Anna Chin

HD
5713.6
.U54
H367
1998

CONTENTS

ACKNOWLEDGMENTS

We gratefully acknowledge the hard work and collaboration of the authors of the case studies: Mulu Birru, Gloria Cross, Joan Fitzgerald, Ann Griffin, Edwin Melendez, John Metzger, Rebecca Morales, David Sweeney, and Leslie Winter. Other senior affiliates of the Economic Development Assistance Consortium who helped out on particular site visits include Jon Gant, Chuck Grigsby, Isabel Hill, Sara Weiss, and Howard Snyder.

A number of student interns and research assistants worked hard on many aspects of this very long project. We thank Charles Abelmann, Meg Barthelt, Pieta Blakely, and Nancy Brune.

We acknowledge the generous support of the staff of the A. Alfred Taubman Center for State and Local Government of the John F. Kennedy School of Government at Harvard University and especially its director, Professor Alan Altshuler, and staff assistant, Kathleen Kaminski, for hosting and providing support for Professor Harrison as a visiting professor during 1994 to 1996.

Our project officers at the sponsoring foundations were intensely involved in the work at many stages. Mark Elliot (then at the Ford Foundation and now at Public/Private Ventures), Bob Giloth at the Annie E. Casey Foundation, and Kavita Ramdas at the MacArthur Foundation were always there when we needed them. In gaining Spreuill White as a program officer at MacArthur (following Kavita's departure), we found ourselves in the unique situation of having the director of one of our case study community organizations suddenly reappear as our banker!

Of all the (very) many kibitzers and other friends of the family who were helpful in building bridges or filling in holes for us, we especially acknowledge Brian Bosworth, Xavier de Souza Briggs, Michele Brookins, Peter Cappelli, Bob Curvin, Ron Ferguson, Marshall Ganz, Steve Glaude, Norm Glickman, Ned Hill, Mark Alan Hughes, Tara Jackson, James Jennings, Langley Keyes, Nolan Lewis, Jr., Susan McElroy, Rick McGahey, Elizabeth Mueller, Dick Murnane, Paul Osterman, Manual Pastor, Jr., Roy Priest, Stu Rosenfeld, Dick Saul, Don Sykes, Walter Thaxton, Florence Williams, Julie Wilson, and Marc A. Weiss (the other). Finally, the last stages of production were greatly facilitated by Janice Sellers-Dunmore at the Milano School.

Thanks to all.

INTRODUCTION

CDC residents, especially those without jobs, need ties to their neighbors for everyday socializing, enhanced feelings of safety, closer monitoring of behavior, and other [personal] reasons. But they also need access to information, job referrals, scholarship recommendations, and other forms of leverage that may call for bridges *outside* the neighborhood. Social capital is not just about "getting by"; it is also about "getting ahead"—gaining access to people and institutions that add information and decision-making clout.

—Briggs, Mueller, and Sullivan (1997, p. 18)

What are employment training (ET) and (more broadly) workforce development (WD) policies and programs for? Obviously, they do not directly create jobs in the short run. They can make a contribution to increasing long-run productivity and therefore to the possibilities for (in President Clinton's famous expression) "growing the economy" over time. If pursued through organic networking with employers—the main theme of this book—they hold out the promise of building relationships of trust and competence that can, over time, reduce the individual and social costs of job search. In the short run, ET and WD policies can relieve temporary or localized shortages of trained workers. Most important, in our view, these programs unapologetically redistribute jobs, earnings, work experience, and dignity to the residents of low-income communities, both urban and rural. In short, they aim to reduce inequality.

Throughout the country, community development corporations (CDCs), other types of community based organizations (CBOs), and such bridging institutions as community colleges have become increasingly involved in boundary-spanning, interregional networks as a way of more effectively engaging in workforce development. In this book, we explain why and how the need has become so acute in America, why conventional approaches to job training continue to fail the persons who need them the most, and—through a series of 10 in-depth case studies—how some CDCs, CBOs, community colleges, and regional quasi-public agencies have directed (or reoriented) themselves to address these concerns.

The Economic Development Assistance Consortium (EDAC), a Boston-based national network of scholars, practitioners, and veteran advocacy lawyers, first learned about these surprising developments in the course of completing a project for the Urban Poverty Program of the Ford Foundation. On a contract originally let in June 1991 and completed in 1993, we produced a progress report that the foundation subsequently published in January 1995. That book, titled *Building Bridges: Community Development Corporations and the World of Employment Training* (Harrison, Weiss, & Gant, 1995), coauthored with Jon Gant, a doctoral candidate in Public Policy at Carnegie Mellon University, reported on the ways in which CBOs around the United States have connected to the evolving ET delivery systems in their cities and regions.

We discovered a number of innovative approaches and experimental projects under way in different locations. The book emphasized that the most effective CBOs—those that have shown themselves to be able to arrange skills training and placement for a nontrivial number of neighborhood residents into jobs paying above poverty-level wages, simultaneously enhancing both participants' sense of self-worth and the reputation of the CBOs—are those that are good at, and assign a high priority to, networking across organizational and territorial boundaries.

We strongly suspect that this observation will not be limited to workforce development and ET activity alone. It probably characterizes many other fields within which community-based groups operate, in their efforts to fight poverty and racial and ethnic discrimination.

The research aroused great interest in CBO, foundation, federal, state, and local government circles. In light of that interest, the Ford Foundation joined forces with the John D. and Catherine T. MacArthur and the Annie E. Casey Foundations and commissioned a follow-up project in which EDAC has been developing 10 in-depth case studies of the evolution, achievements, and problems encountered by three different kinds of workforce development and ET networks.

We developed several objectives for ourselves and our clients. Most immediate, we wanted to tell the stories of these innovative, energetic community-based (or community-minded) organizations in all of the detail and depth that they deserve. There are valuable lessons to be learned from studying even those groups and projects that, by their own standards, have not been especially successful at interorganizational, boundary-spanning network creation or management. Of course, the most important "audience" for these stories is the huge pool of similar organizations throughout the country that may be contemplating network approaches to workforce (and other aspects of community, urban, and regional economic) development.

Although the current research does not constitute formal program evaluation in the conventional sense, we think we have learned much that will be of value to those charged with assessing the extent to which participation in interorganizational, cross-boundary networks significantly enhances the goals of a CBO's workforce development activities—relative to what might have reasonably been expected in the absence of such networking approaches. Thus, we give particular attention to identifying process and outcome performance measures that are in principle knowable and even quantifiable—by evaluators, funders, public sector agencies, and the public policy community at large.

What do we mean by **network**? The economist Chris Tilly (1996) has come up with a complex but exceedingly rich conception of how various kinds of networks, or webs, relate to one another, and how they fit into the larger context of the structure of work itself. He writes,

Although solitary workers certainly exist, in general work depends on **transactions** among parties, notably between products and immediate recipients of use value added by work. Transactions consist of interpersonal transfers of information and/or goods of which the parties are aware; they become work transactions when the effort of at least one party adds value to the element transferred. Organized, durable transactions cluster into implicit or explicit **work contracts** stipulating parties, rights, obligations, and sanctions to the transactions in question. Work contracts differ from mere accumulations of work transactions in featuring enforceable agreements that govern durations, limits, enforcement mechanisms, and relations among transactions.

Work contracts are typically embedded within a variety of social networks, including production networks. **Social networks** in general are sets of relations among persons, organizations, communities, or other social units. Most commonly we think of a single network as the set of relations among specified actors of a given type defined by a certain kind of tie: for example, the interlocking of corporate boards of directors or shared involvement among supporters of a social movement. A **production network** consists simply of a connected set of work contracts linking multiple producers and recipients. **The social structure of work** centers on concatenated contracts: Jobs, occupations, careers, firms, unions, labor markets, discrimination, and inequality all appear as special cases or outcomes of that concatenation. (p. 3)

Webs of interpersonal and interorganizational relationships come in many sizes and shapes. Our main interest in this book will be in those webs that ultimately connect the residents of low-income neighborhoods to employers with real training positions or—better yet—job vacancies via such mediating institutions as community colleges, trade associations, government agencies, and especially CDCs or other CBOs.

We ought not to be surprised that organizations based in inner-city neighborhoods have connections to the larger cities and regions of which they are a component. Yes, mainstream social science and the typical media treatment have for many years tended to depict racial or ethnic "ghettos," and the people who live within them, as isolated, socially disorganized, and lacking in institutional richness. This depiction, however, is at best an overgeneralization—even a stereotype—and at worst factually incorrect. As sociologist Melvin Oliver (1988) has demonstrated empirically for Los Angeles, the interpersonal networks to which the residents of even the poorest and geographically most seemingly disconnected places such as Watts have access are actually quite spatially dispersed. University of Chicago public policy professor Richard Taub (Taub, Surgeon, Lindholm, Betts Otti, & Bridges, 1977) was surely right to urge us—as long ago as the mid-1970s—to (in Oliver's paraphrase) "study the structure of ties without regard to spatial borders."

A second technical concept in need of a working definition is **workforce development**. This term is commonly treated (incorrectly) as a synonym for "job training." Instead, we understand workforce development to consist of a constellation of activities from orientation to the work world, recruiting, placement, and mentoring to follow-up counseling and crisis intervention (Harrison et al., 1995). The actual training is but one element.

Still another technical term now much in use by funders, and that appears often in our case studies, is **sector**. In their superb analysis of privately initiated sectoral community development strategies (with **sectoral** referring to efforts focused on one or a few specific occupations or narrowly defined, closely interrelated industries that use common technologies or draw on similar resources such as particular occupations), Peggy Clark, Steve Dawson, and their colleagues (1995, p. viii) distinguish four stages of organizational and project development. The highlights of their astute research apply to all the community-based but regionally engaged networks that we have been tracking for as long as 5 years.

New initiatives need especially to decide what is for them an appropriate mix between formal and informal analysis of possible projects. Even as they enhance the labor market chances of low-income persons, emerging initiatives also need to identify and highlight intended noneconomic outcomes, such as involving a larger number of neighborhood residents in decision making or enhancing the reputation of the organization. Operators of mature initiatives face two enormous challenges: remaining focused but also being continually creative. Expansive initiatives must (as their name implies) figure out how to sustain their organizational goals, objectives, and integrity as they "move to scale."

In research and policy analysis on local and regional workforce and economic development, the terms *sector* and *cluster* are sometimes treated synonymously. We prefer Brian Bosworth's (1996, p. 54) distinction, especially his conception of cluster as a geographically bounded concentration of similar, related, or complementary businesses with active channels for business transactions, communications, and dialogue that share specialized infrastructure, labor markets, and services that are faced with common opportunities and threats.

CBOs should try to identify and connect to such clusters rather than (or in addition to) identifying individual potential partner firms and agencies, if only because many companies hire from their suppliers, vendors, customers, and collaborators in the design of new equipment and products. Relatively low-skill-level employees, as well as engineers, financial analysts, and other white-collar occupations, are increasingly finding that their best opportunities for upward mobility lie in moving among firms and agencies within a cluster as the chances of moving up within a particular employer (let alone achieving "lifetime employment") fade into history.[1]

Probably the single most important conclusion from our research is that networking per se ought not to be seen as a substitute for acquiring organizational capabilities but rather as a stimulant or complement to them. We have learned that effective networking does not just happen by joining a consortium. It requires strategic planning to choose the most appropriate networks to create true win-win partnerships. Once the appropriate networks have been chosen, an institution must be prepared to commit considerable resources to reap any benefits. Networking typically takes a sustained commitment before benefits can be realized.

Perhaps equally important, having network connections is not enough. It depends on what kinds of connections and how the connections are institutionalized. For example, a detailed statistical analysis of interview data on approximately 1,800 Boston and eastern Massachusetts households in 1993 and 1994 has been conducted as part of the Russell Sage and Ford Foundation-sponsored Multi-City Study of Urban Inequality. Two University of Massachusetts researchers associated with the Boston Urban Inequality Research Group, Edwin Melendez and Louis Falcon, found that three of four recent migrants to the region had close network connections to people living there—what sociologist

Mark Granovetter (1985) calls "strong ties." This was especially true for Latinos. Employment found through relatives, friends, or acquaintances, however, tended to connect black workers to jobs paying lower wages and benefits and characterized by lower socioeconomic status. Jobs obtained through these kinds of networks provided fewer and lower benefits for both blacks and Latinos and tended to be situated within workplaces (or in occupations within workplaces) that were disproportionately black or brown—in other words, relatively segregated. By contrast, networks of this kind seem to make no difference one way or another to the kinds or quality of jobs obtained by non-Hispanic whites. In short, it is possible to be well connected—but to the wrong kinds of networks (Falcon & Melendez, 1996)!

Two other introductory comments are in order. In this field, as in many others, there is perpetual disagreement and confusion over whether individual "heroes"—especially the "charismatic leaders"—or well-ingrained institutions hold the key to shaping events. Veteran organizer and political sociology theorist Marshall Ganz (1995) developed a brilliant and original idea that incorporates both into a dynamic, historically grounded theory of social change. We present his model in greater detail in connection with the summary of our case study of the San Jose-based, largely Chicano, Center for Employment Training.

For now, suffice to say that institutions and their rules do tend to shape and constrain the behavior of individual actors—in "settled times." In "unsettled times," however, when the prevailing rules of the game are being widely challenged, particular actors can take the lead in transforming institutions. It is those actors who are least wedded to the status quo and who belong to many intersecting social networks without being central to any one of them—the so-called "borderlands" actors—who are likely to be the most effective in leading institutional change. Some (not many) CBOs fit this description. Other groups that already fit it well might be sought out and actively brought into the workforce development movement.

Finally, much of the theory of organizational networks that informs this research came originally from the study of corporate strategy and regional economic development. There will be plenty of opportunity throughout the book to review the relevant literature. An important theme that has emerged from these fields during the past 12 years is that firms must learn to strategically cooperate as well as compete with rivals. This is a key principle underlying the best of the community-based workforce development networks we have been studying. We provide many examples throughout the book.

The need for new organizational approaches to employment training and workforce development is compelled by extraordinary changes that are under way in how labor markets function, especially regarding what economists call the "demand side"—that is, in the recruiting, hiring, training, and pay policies of employers. Therefore, an overview of those changes is provided in Chapter 2.

NOTE

1. For a superb analysis of the regional labor market aspects of an industrial cluster—in this case, the high-tech microelectronics, software, and venture capital companies that make up Silicon Valley—see Saxenian (1994).

2

THE CHANGING STRUCTURE
OF LABOR MARKETS IN
AMERICAN CITIES

American prosperity is extremely uneven. Families and workers at the
top of the economic ladder enjoy rising incomes. Families in the middle
have seen their incomes stagnate or slip. Young families and workers at
the bottom have suffered the equivalent of an economic depression.
Though the nation is in the midst of an economic expansion, recent
Census statistics offer little hint that the trend toward wider inequality
has slowed. So long as the trend continues, a large minority—perhaps
even a majority—of Americans will believe that prosperity has passed
them by.

—Burtless (1996, p. 24)

The last national recession bottomed out during the first 3 months
of 1991. For an entire year thereafter, there was no job growth in
the country at all as employers hedged their bets, waiting to see what
would happen—meanwhile extracting extra production from existing
employees. This was the period of what the media called the "jobless
recovery." Had it continued, this really would have been a new development in the history of American labor markets.

By the spring of 1992, however, jobs were again being created—at
approximately the average rate that had prevailed among all eight

previous business cycle recoveries dating to the end of World War II. Even including the year of "jobless growth," the U.S. economy created more than 11 million new jobs between the spring of 1991 and the fall of 1996.

Contrary to many casual impressions—and not a few analyses by the experts and in journalistic commentaries—the U.S. Bureau of Labor Statistics (BLS) makes it quite clear that, if both new jobs created and hires to replace vacancies in "old" jobs are counted, relatively low-skilled positions (defined by the BLS as "jobs that can be learned quickly and that generally do not require post-secondary education") accounted for most of the wage and salary employment created in the United States between 1983 and 1993 (Rosenthal, 1995). Moreover, projections based on the BLS's widely used *Occupational Outlook Handbook* (n.d.) conclude that, for the future period 1992 to 2005, again considering both new jobs and replacement vacancies, only approximately one of eight higher than average growth occupations will require a college degree, whereas fully two thirds will require no more than a high school diploma (Mangum, 1995). It is quite amazing how powerfully the myth of a steadily disappearing demand for low-skilled labor in America has been perpetuated.

None of this implies that education is unimportant—that would be foolish. Nor is it inconsistent with Harvard University economist Ronald Ferguson's (1996) finding that racial differences in basic math and language skills (as measured by standard written tests) are strongly correlated with the racial wage gap. It does indicate, however, that the alleged disappearance of low-skilled job opportunities in America has been exaggerated. There is, and will continue to be, considerable room in the economy for workers with modest formal schooling. The main problem is to dramatically improve the quality and reliability of basic skills for the majority of youth who will not go beyond high school. They, along with adults undergoing retraining, are increasingly expected by employers to be better equipped to learn new skills on the job, to take further training, and to be, by mainstream standards, willing and able to accept the disciplinary requirements of most workplaces.

In an institutionally racist society, of course, the continued creation of jobs with modest skill requirements does not guarantee equal access of all prospective workers to even these "low-end" jobs. In the big cities especially, a new wave of ethnographic scholarship—the scholarship of direct engagement, not relying solely on the manipulation of census data—has documented a complex and intense competition for these jobs among men and women, residents and newcomers, and (among the latter) between those who have traversed very different paths (experienced different modes of "insertion") into the urban economy (e.g., Morales & Bonilla, 1993; Sassen, 1989; Thompson, 1997; Waldinger, 1996; Waquant, 1994). In this competition for low-end jobs, racial divisions and years of residence in the city go far toward explaining who holds which places in the system.

The upshot is that the signal problems that plague urban labor markets—and employment systems everywhere—consist mainly of low wages, job insecurity, and inadequate opportunities for on-the-job learning. They are not the result of some technologically (or otherwise) driven "twist" toward employers' greater need for high skills per se. Barbara Bergman's old "crowding" hypothesis,[1] in the context of greatly weakened labor regulation (enforcement of national wage and hour and safety and health standards), seems a far more useful way to think about what is happening than the hopelessly technologically determinist myths about labor market restructuring that have come to dominate mainstream discourse, especially within academic economics.

SKILL, EARNINGS, RACE, AND SECTORAL SHIFTS IN THE URBAN ECONOMY

The categories "skill," "schooling," and "ability to learn" require careful, nuanced distinctions. A new survey of approximately 3,000 employers in four metro areas (Los Angeles, Detroit, Boston, and Atlanta), conducted by Harry Holzer (1996), reflects the importance of recognizing these distinctions. Holzer reports that the extent to which employers require high school diplomas, specific experience, or prior training and references—but not post-secondary education—as a condition for hiring into entry-level jobs that include tasks such as daily reading and writing, arithmetic, use of computers, or dealing with customers increases the probabilities that blacks and women will be hired (vis-à-vis white men) and the level of starting hourly wage rates.[2]

A recent national survey of hiring, training, and management practices in a representative sample of (by coincidence) 3,000 companies—
the first official survey of its kind—was conducted by the Census Bureau
in August and September of 1994. This was done for the National Center
on the Educational Quality of the Workforce and the Institute for
Research on Higher Education, both based at the University of Pennsylvania. The survey results show that employers place a higher priority on
"attitude," "communication skills," "previous work experience," "recommendations from current employees," "recommendation from previous employer," and "industry-based credentials certifying skills" than
they do on years of schooling, test scores, grades, or teacher recommendations.

At the same time, managers in small focus groups told center codi-
rector and institute director Robert Zemsky that they were extremely
skeptical of the quality of most young graduates of the nation's schools
(Applebome, 1995). What employers say they want is people with better,
more reliable schooling in job-relevant skills, not necessarily people with
more schooling, per se.[3]

Thus, the economy continues to generate job opportunities both
through the creation of new employment and via turnover in existing
slots. Also, employers are demanding new workers with better basic
skills but hardly with rocket science under their belts (or in their heads;
the typical level of "computer literacy" that employers demand can often
be taught in one solid semester in a high school or community college
computer lab). The problem is that average wages and benefits have
continued to decline, even though average productivity and profits are
growing. This is by now a well-researched subject, with substantial
agreement on the "facts" (if not on the relative importance of the
causes). Figure 2.1 displays the pattern of inflation-adjusted hourly
wages since 1973 by education. Clearly, it is those with the fewest years
of schooling—precisely the segment of the population that is disproportionately concentrated in the inner city and in poor rural areas—who
have lost the most ground. During the 1990s, however, even those with
a 4-year college degree have seen their real earnings growth flatten
(Mishel & Bernstein, 1994, pp. 140-145, 364-366). Women working
full-time experienced rising median real weekly earnings during most
of the past 20 years (even as the average earnings of men fell steadily),

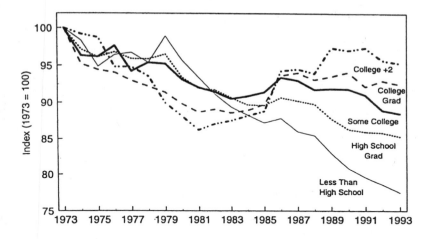

Figure 2.1. Hourly Wages by Education, 1973 to 1993
SOURCE: From Mishel and Bernstein, 1994, p. 141

but that trend came to an abrupt end in 1994. Since then, women's earnings have also been falling (Mishel, 1995, p. 61).

The government keeps statistics from the vantage point of employers as well. The BLS's "employment cost index" measures spending by companies, government agencies, and nonprofit organizations on pay and benefits for each hour worked. Figure 2.2 shows percentage changes between each quarter of a year and the corresponding quarters 1 year earlier. All civilian employees are included. "Benefits" include employer expenses for health insurance, vacations, holidays, sick leave, shift premia, life insurance, pensions, unemployment insurance, and severance pay. The annual growth rates of both wage and benefit costs to employers (per hour of work) show strong declining trends since at least the early 1980s (with the predictable cyclical bump up associated with tighter labor markets prior to the 1990-1991 recession). These figures are not adjusted for inflation. After accounting for increases in the cost of living, wage and salary growth by 1993 was actually falling behind changes in the cost of living (Hershey, 1995).

Of course, what to managers appears as "success" in containing the growth of contributions to their employees' health insurance premiums

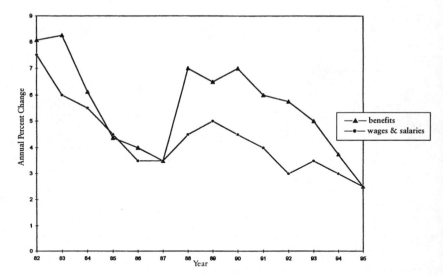

Figure 2.2. Nominal Hourly Employment Cost Index, 1982 to 1995
SOURCE: From U.S. Bureau of Labor Statistics [Reprinted in *New York Times,* November 1, 1995, p. 2]

and other benefits spells growing hardship for a rising share of the population. The already inadequate contribution of the government to social insurance—especially to welfare for poor relief—was effectively compromised in 1996 and is likely to erode further in the years immediately ahead.

As for urban communities of color, for many years it was true that African Americans, in particular, depended on and benefited from a healthy, growing public sector for jobs and for social services of all kinds.[4] Black men and women employed in government average significantly higher wages than do African Americans who are employed in the private sector (Table 2.1). A much larger fraction of them are likely to be employed year-round and full-time as well. The advantage of working for the government is especially great for women. Actually, what the Census Bureau data in Table 2.1 reveal is that public-sector employment is the great equalizer in American working life. For blacks and other workers of color, the likely decline in the government's long-standing one-sixth share of total employment in the country under any

Table 2.1 Median Annual Earnings of Year-Round, Full-Time
(YRFT) U.S. Workers: Comparison of Public
Employees to All Workers by Race and Sex—1989

	Government Sector		All Sectors	
	Median Annual Earnings ($)	*% Working YRFT*	*Median Annual Earnings ($)*	*% Working YRFT*
Men				
Black[a]	26,626	80	21,971	64
All races[b]	31,274	85	26,269	66
Women				
Black[a]	21,346	73	18,036	57
All races[b]	21,760	75	18,351	52

NOTES: a. 1990 Census of Population, *Characteristics of the Black Population*, 1990 CP-3-6, Tables 46-47; refers to earnings in 1989.
b. Current Population Reports, Consumer Income, Series P-60, No. 180, *Money Income of Households, Families, and Persons in the United States: 1991*, Table 33; deflated back to 1989 for purposes of comparison using the Consumer Price Index change of 12% inflation.

of the "balanced budget" austerity regimes now being debated in Washington is bad news, indeed.

A second issue concerns the current wave of mergers and consolidations among banks, hospitals, and other typically urbanized service-sector firms. These are the employers that have been the least likely to shut down existing operations to relocate to the suburbs (let alone overseas), choosing instead to set up additional branches. They, too, have for many years been important sources of employment opportunities for workers of color ("health," in particular, appears in most analyses of "good bets" for targeted economic development and job training programs). They will continue to be important—especially the health sector, given the graying of the population. Moreover, as the big hospitals continue to downsize and outsource services to smaller independent contractors, nursing care facilities, and other organizations under the cost-cutting aegis of "managed care," jobs will be created in these "outside" workplaces. Still, the extent of the growth of this sector as a whole will depend in part on political decisions about public spending, regulation, and the pace of deregulation in the years ahead.

Whatever the extent of the slowdown, the burden is likely to fall disproportionately on the residents of low-income communities of color in terms of both services forgone and jobs and wages lost.

In summary, job opportunities are being created, and the great majority of them do not (and, in the years just ahead, probably will not) require a college education. Those sectors, however, that have been especially important sources of employment for the residents of largely black, Latino, and other urban neighborhoods—hospitals and other health facilities, government, and (recently) banks—are no longer creating jobs at a pace commensurate with their track record of even the recent past. Finally, average wages and even benefits (or, by another measure, their rates of growth) continue to decline, year after year, for a large fraction —by some accounts, perhaps four fifths—of the population.[5]

DECLINING JOB SECURITY AND
DIMINISHING RETURNS TO EXPERIENCE AND SENIORITY:
THE DEVOLUTION OF INTERNAL LABOR MARKETS

Far-ranging changes are under way in how companies organize work: how they produce, where they locate the work to be done, and whom they hire to do the work. A central feature of these changes is what now appears to be a long-run ("secular" as opposed to periodic or "cyclical") move away from what—at least for the most profitable operations of the biggest, most visible, and influential firms and agencies—constituted the dominant (if never universal) employment system of the post-World War II era.

In the three decades or so after 1945, with job tasks substantially standardized and broken down into a finely grained division of labor, managers commonly hired new young workers at the bottom of job ladders, trained them on the job, promoted people within the organization, and paid wages more in accord with seniority and experience than with individual productivity or current firm performance. Although few companies ever formally committed themselves to literally "lifetime" employment, there was widespread expectation that (at least for white men, but gradually for others as well) there would be a high degree of long-term job security, with occasional disruptions in work triggering the receipt of unemployment insurance and other forms of temporary income maintenance. Seniority systems were promoted by unions, and within the civil service, as a means of achieving fairness. Business came to support them because, by according greater job security to more experienced employees, the latter could better be counted on to train and otherwise teach skills and know-how to younger workers within the organization. The question of "who is poor" or underemployed amounted importantly to understanding what tracking mechanisms in the society systematically blocked which groups from having access to jobs in these "internal labor markets."[6]

Whether there is a well-defined, coherent new "system" of work coming into existence—accommodating to globalization, deregulation, the shortened shelf lives of products, the enhanced capability of an increasing number of competitors to quickly erode the advantages of (play catch-up with) the innovators through "reverse engineering," and generally heightened competition—is still an open question. There is a growing consensus among scholars, however, that the old system is coming apart. This is not the place to examine all the aspects of this "devolution" of internal labor markets, the growth of "flexible" forms of work organization, and the implications for training and economic development.[7] But we can at least sketch the main features of what a number of researchers believe is happening to how work and labor markets are being reorganized.[8]

The changes are motivated by a complex of reasons: deregulation, greater actual or potential competition from abroad, growing numbers of corporate hostile takeover attempts, and other signals from stock-holders that put a premium on short-term firm performance. All this has made managers increasingly conscious of short-term fixed (or, as economists like to describe the wages of workers on long-term implicit contracts, "quasi-fixed") costs and committed to reducing them whenever and as much as possible. First with IBM, then with Xerox, and recently with AT&T and the big banks, the stock market has instantly rewarded companies that cut costs through consolidations, mass layoffs, and wage and benefit rollbacks by bidding up share values, which only further encourages this kind of management behavior. Even as progressive voices—from the former U.S. Secretaries of Labor Ray Marshall and Robert Reich to labor economist Richard Freeman and political theorist and activist Joel Rogers—were advocating that more companies should take (and stick to) the "high road" in labor policy, the business community was organizing itself (e.g., through the Labor Policy Association) to advocate for, and legitimate, greater "flexibility" and management discretion in work arrangements.

What this amounts to, in practice, is a proliferation of different forms of work organization, blurring the traditional distinctions between "core" and "periphery," "permanent" and "contingent," "inside" and "outside" employees, and "primary" and "secondary" labor markets. Thus, managers employ some workers in more or less routine wage and salary positions inside the firm or agency. But they also hire temporary help agencies, management consulting firms, and other contractors, to provide employees (ranging from specialized computer programmers to janitors and clerical personnel) to work alongside the "regular" people but on short-term assignments and under the management of the contractor. Companies, colleges, and hospitals outsource work that was formerly performed in-house to outside suppliers in the United States and abroad. They also shift some work from full-time to part-time schedules, in part to avoid federal labor regulations covering wages and benefits that have been interpreted by the courts as not covering "leased" workers. The "temp" agencies and other contractors are being used increasingly by managers in the "focal" firm as a mechanism for screening potential regular employees, with candidates serving their "probation" on the payrolls of the contractor before moving into the focal company or agency. This creates the possibility for further inequities because two persons working side by side for an outside contractor may be equally competent, but one will eventually be absorbed into a full-time job, whereas the other will continue to bounce from one temporary assignment to another.

Two concrete expressions of this growing heterogeneity in work organization and practices, both among and within particular employers, are declining employment security and more uncertain wage and salary prospects over time. Surveys conducted by the American Management Association show that managers increasingly regard layoffs as "strategic or structural in nature" rather than a response to short-term, temporary business conditions (cited in Cappelli, 1995, p. 577).[9] The fraction of laid-off workers who can expect recall into their old jobs has

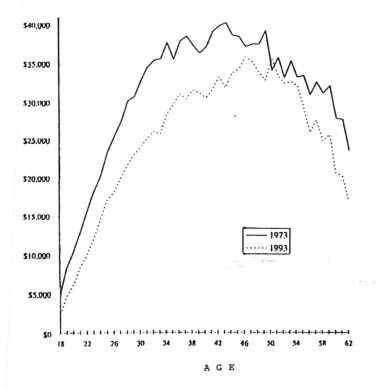

Figure 2.3. Mean Real Annual Earnings of All 18- to 62-Year-Old Males, by Age, 1973 and 1993 in 1993 Dollars
SOURCE: From Pines, Mangum, & Spring, 1995, p. 17

been lower in the post-1991 recovery than in the previous four national business cycles, whereas those who do get reemployed somewhere are more likely than before to land part-time jobs (Cappelli, 1995, p. 578).

The flatter organizational structures—another workplace "innovation" whose introduction in recent years has been responsible for many of the layoffs of middle-level managers—contribute to reduced promotion opportunities within the surviving internal labor markets of companies and agencies. Generally, the payoff to seniority, as measured by age-earnings profiles, is shrinking over time both across the workforce (Figure 2.3) and, more precisely, over the careers of specific individuals, including those who stay with the same employer.[10] Also, compared to expectations formed in earlier periods (such as the early 1970s) or to the average earnings profile for their age group, workers in the 1980s and 1990s face increasingly fluctuating earnings from one year to the next.[11] Quite apart from questions of "fairness," Cappelli (1995) and others question the long-run implications of this growing instability in earnings streams for the continued smooth functioning of a consumer economy built on households borrowing against future income and having to make regular payments over time for items ranging from housing and cars to their children's education.

In summary, the mix of types of jobs and work schedules is becoming more diverse. In Cappelli's (1995, p. 570) careful and sober formulation, "there is a continuum between 'pure' internalized arrangements and complete market determination along which employers are moving, and the argument here is that, on average, practices are shifting along that continuum" from the former in the direction of the latter. It would be one thing if this were mostly a voluntary development serving the needs of a greater share of the population. All serious researchers agree, however, that these changes are being initiated mainly on the "demand side" of the labor market, with managers seeking to reduce their exposure to long-term fixed obligations. It appears that a growing fraction of pay is becoming "contingent"—on individual job performance, on the fortunes of the employer, on the current state of the "animal spirits" in the stock market, or on what the company thinks it can get for its money by turning to suppliers in other, lower-cost locations.

NOTES

1. For the latest application of the "crowding hypothesis" that she first developed in the early 1970s, see Bergmann (1996).

2. Holzer also believes, however, that formal testing by employers generally works against black men because they are less likely to pass the tests. Chris Tilly and colleagues think that, on balance, formal screening and hiring procedures (although not necessarily testing per se) help black men, whereas informal methods (which are probably on the rise because of the reduced enforcement of affirmative action and related regulations by government) allow more room for employers to arbitrarily discriminate (Kirschenman & Neckerman, 1991; Moss & Tilly, 1996b).

3. The results of Zemsky's interviews are reported in Zemsky (1994). Mainstream approaches to the apparently rising "returns to skill" are abundant. One of the best is provided by Freeman and Katz (1994). For some thoughtful alternative interpretations of the same data, see Howell (1994, 1997), Levy and Murnane (1992), and Mishel and Teixera (1991).

4. The most comprehensive statistical documentation and thorough literature review on the jobs aspect of this assertion is contained in a working paper by Michael Leo Owens (1996).

5. By late 1996, under the pressure of what, these days, are relatively low rates of unemployment, there have been signs of small increases in real average hourly wages (although even this claim is contested; what is happening to average wages depends on which data sources one consults). No improvement in the two decade-long increase in earnings inequality is in sight, however. On changes in the distributions of family incomes, wealth, individual labor market earnings, and benefits, the most comprehensive survey is by Frank Levy (1995). For especially accessible treatments of particular aspects of what Lester Thurow once called "the surge in inequality," see Bluestone (1995), Krugman (1992), and Wolff (1995).

6. For a sampling of classic and enriched views of the postwar system of internal labor markets and their implications for employment training, racial discrimination, and urban poverty, see Berger and Piore (1980), Doeringer and Piore (1971/1985), Gordon (1971), Gordon, Edwards, and Reich (1982), Harrison (1972, especially Chap. 5), Jacoby (1985), Osterman (1984, 1988), and Peck (1996).

7. For in-depth explorations of these questions, see Appelbaum and Batt (1994), Bailey and Bernhardt (1996), Cappelli (1995), Cappelli et al. (1997), Harrison (1994), Osterman (1994, 1995), Pfeffer and Baron (1988), and Piore and Sabel (1984).

8. The following draws especially heavily on the recent writing of Peter Cappelli. There and elsewhere, readers can learn about other new developments, from the growth of teams to the breakdown of narrow job descriptions and the enlargement of many existing jobs, with those who do have work being required or enabled to perform a greater variety of tasks, often using productivity-enhanced new technology. The changes that are occurring in the organization of work in America are not all "bad"—provided one can break into the system.

9. Such permanent layoffs ("displacements") are currently actually higher for managers than for other occupations, after other characteristics are taken into account.

10. The flattening of age-earnings profiles reflected in Figure 2.3 is suggestive, but longitudinal studies on specific workers or on cohorts are more definitive. For such evidence, see Marcotte (1994), Bluestone and Rose (1997), and Rose (1995). A new study from the Employee Benefit Research Institute reports that between 1991 and 1996, median job tenure for men aged 25 to 64 fell by nearly 20% (see "Economic Trends," 1997).

11. A technically sophisticated study is Gottschalk and Moffitt (1994). Some of this is due to the growing incidence of performance-based pay and bonuses and, as such, is built into the new work systems by design. Some is also probably attributable to declining job tenure, although (as Cappelli [1995] explains in footnote 10) "tenure" is a confusing indicator of job security because it is affected by changes in both quits (mostly voluntary) and layoffs (usually involuntary). Unfortunately, the Census Bureau stopped collecting data distinguishing between quits and layoffs in 1981, at the beginning of the Reagan administration.

3

TAKING STOCK OF WHAT WE KNOW ABOUT JOB TRAINING AND WORKFORCE DEVELOPMENT

We have always tried to spread too little money over too many adult enrollees, after short training periods, recycling people back into the low wage jobs many like them were getting without enrollment; and we have attempted to remedy the problems of disadvantaged youth after the damage was done by enrolling them in a variety of remedial programs on a ratio of approximately 1 week to each past year of their . . . troubled lives. . . . On-the-job training has always paid off better than classroom [training] but employers have [generally] been reluctant to get involved.

—G. Mangum (1995, p. 4)

Although wages, benefits, and opportunities for secure long-term employment in American labor markets are all becoming buffeted by forces far beyond the control of any individual job-seeker, new jobs and job openings are being created, and at about the average pace of the past half century. The pace of job creation continues to be inadequate to provide employment for all who seek it. Jobs are, however, being created—in the millions. That in itself would make it compelling for community-based organizations (CBOs) to have an interest in access to quality employment training (ET) for the residents of their neighborhoods.

In fact, the case for such interest is even stronger. This is because workforce development—the constellation of activities from recruiting, placement, and mentoring to follow-up, of which the actual training is but one element—not only involves the "production" of skills but also enhances trainees' ability to learn and socializes them to working with others in settings defined by managers in private firms and public or nonprofit agencies. Employers select job applicants based on the employers' perceptions of all these. Indeed, for entry-level jobs, Ronald Mincy, Chris Tilly, and others have argued that "skill," in the narrow, technical sense, may be less important to many gatekeepers than perceived attitudes and socialization—what have come to be known as the "soft" skills (Moss & Tilly, 1996a; Pouncy & Mincy, 1995). This has implications for hiring discrimination, to the extent that—as Tilly and others believe—employers are more easily able (and likely) to discriminate against black men when screening and hiring are done more informally and subjectively. The soft skills are surely more likely to be "processed" informally than through actual tests or by certification (Moss & Tilly, 1996b). By this reasoning, it is probably getting easier for employers to screen out black men whom they do not want to hire for reasons other than lack of skill per se.

Obviously, training without economic development will always leave some people unemployed or underemployed at poverty wages; in the game of musical chairs, everyone does not get to sit down if there are not enough seats. For any given supply (or rate of growth) of job vacancies (i.e., demand for labor), however, those who have more (and higher quality or more unambiguously certified) training tend to be the ones most likely to get a seat. As will be shown later, this may be due as much to the reputation and "connectedness" of the training organization as to the superiority of the training "model" per se.

WHAT KINDS OF TRAINING
WORK BEST AND FOR WHOM?

The most recent bevy of evaluations, surveys, and metasurveys on the costs and benefits of ET in particular, and of workforce development, more generally, substantially replicate findings that have been well-known to researchers and policymakers in the field for many years. In other words, with regard to the question, "What works and what doesn't?" (the title of a recent report from the U.S. Department of Labor, 1995), surprisingly little has changed over time.

The most general conclusion to be drawn from this substantial literature is that for adults, there are "serious doubts about the effectiveness of stand-alone basic education and GED programs" (Savner, 1996, p. 5). Training sponsored or conducted directly by employers generates relatively greater benefits, in terms of wage improvements or reduced chances of early subsequent unemployment, than does any other form of training. Vouchers provided to adults to return to school so they can then search for jobs on their own have proven to be especially ineffective.

The earnings of young people still in school—especially young women of all races and ethnicities—are on average enhanced by occupational training in high schools, community colleges, and (some) vocational and technical organizations. By contrast, the payoff to school-based training for young African American men in recent years has been essentially zero (this is not true for Armed Forces training, whose universally acknowledged effectiveness may provide a clue to what is going on—what scholars will call "reputation effects").

The group for which practically none of the existing approaches have worked is out-of-school youth. There are a handful of exceptions, notably, the Job Corps and the San Jose-based, largely Chicano, Center for Employment Training (CET). We will later characterize this organization as the single most innovative, and demonstrably successful, nongovernmental training apparatus for poor people in the United States. Short-term (3-6 months) classroom training is generally not very effective for any group, with CET being the remarkable exception to the rule.

Cornell Professor John Bishop, former Department of Labor Chief Economist Lisa Lynch, Massachusetts Institute of Technology's Paul Osterman, and other specialists have amply demonstrated that employer-centered training produces the greatest payoffs (Bishop, 1994a, 1994b; Knocke & Kalleberg, 1994; Osterman & Batt, 1993). Indeed, Bishop has shown that employers themselves clearly benefit from such investments through increased worker productivity and reduced turnover.

CONTINUING SYSTEMATIC UNDERINVESTMENT
IN YOUNG PEOPLE BY THE PRIVATE SECTOR

Company-based (or closely linked) training may show the best results, but—especially for entry- and other lower-level, generally younger, employees—it is hard to find. Although precise estimates vary, the general conclusion that U.S. firms systematically underinvest in firm-specific training, especially for their less-skilled employees, appears in many studies. According to Lynch (1993), the U.S. private sector underinvests in firm-specific training compared to private firms in Japan, Australia, France, Germany, The Netherlands, Norway, Sweden, and even the United Kingdom.[1]

As for the bias against spending on youth (and on low-skilled workers, generally), Ray Marshall and Marc Tucker (1992) claim that at least two thirds of the annual spending of all U.S. companies on education and training during the 1980s was invested in their college-educated professional employees. Since the early 1980s, only 4% of non-college students aged 16 to 25 have received formal company-centered training of at least 4 weeks duration (Lynch, 1993). Between 1983 and 1991, the average duration of formal training paid for and provided by employers for learning a new job declined substantially, especially for workers with less than 10 years of seniority (Cappelli, 1995, p. 571). Moreover, although recent surveys show that the majority of private firms do make some investments in worker training, the private-sector American Society for Training and Development reports that a tiny handful are responsible for the vast majority of the total investment—one half of 1% of all firms spend 90% of the total (Henckoff, 1993, p. 62).

The traditional explanation for this investment behavior on the part of private companies, in the context of American political institutions, is that workers in the United States are more likely than those elsewhere to take whatever training they receive from a particular firm and "jump ship" to another. At the level of high theory, Nobel economist Gary Becker many years ago drew a powerful distinction between firm-specific and more "general" training, predicting that private businesses would always underinvest in the latter because it could be used by footloose employees to ultimately benefit competitors. Bishop (1994a,

p. 25) confirms that, during the past 20 years, a smaller fraction of American workers have been on their current job (with their current employer) for more than 5 years than almost anywhere else; the rates for Japan and Germany were half again as great. This nexus forms the basis for economists' widely shared classification of general ("portable") training as a "public good" that, if not subsidized by government (however and in whatever settings it is actually provided), will be systematically underproduced.[2]

At an empirical level, part of the explanation for the relatively low average flow of private spending on employer-centered training is that most firms are small and the incidence of spending on training declines dramatically the smaller the establishment (Bishop, 1994a, p. 8).[3] The reasons are apparent: Small firms have smaller budgets, less organizational slack for managing such activities, and—partly because they also on average pay lower wages and benefits—confront higher probabilities that employees will "job hop" sooner. As with medical care, occupational health and safety, environmental concerns, and access to information about new technologies, it is precisely the smaller business sector of the economy that is most in need of a combination of the carrot of targeted public policies and the stick of regulatory sanctions if we are to maintain a higher overall standard of living.

There are, of course, numerous examples of private companies (generally, among the largest ones) that have made major commitments to firm-based training. Motorola, Xerox, Corning, General Electric, Hewlett-Packard, Ford, Federal Express, AT&T, and Siemens are among the companies that are most frequently cited for their relatively expansive approaches to employee training (Henckoff, 1993). Whether these "success stories" add up to a trend toward restructuring corporate environments as "high-performance workplaces" is, however, still contested. Even many of the early advocates, such as Appelbaum, Bluestone, and Cappelli, must be said to have become "cautiously pessimistic."

Unions have also (and traditionally) played an important role in negotiating to get employers to make investments in training that they would not have made otherwise. A good example is afforded by AT&T. Under a contract dating back to 1986, the firm, the Communications Workers of America, and the International Brotherhood of Electrical Workers established a joint labor-management training fund to support training of workers over and above what the company normally provides that is targeted especially to help employees meet the entry-level requirements for changing jobs within the firm or for preparing for new employment outside AT&T. Similar funds exist in the automobile industry, notably between General Motors and the United Auto Workers. The durability of the AT&T commitment is currently being tested as never before, given the mass restructuring-related layoffs now under way in that company.

PROSPECTS FOR CLOSER PARTNERSHIPS BETWEEN PRIVATE COMPANIES AND COMMUNITY COLLEGES AND THE IMPLICATIONS FOR THE TRAINING AND SUCCESSFUL JOB SEARCH OF RESIDENTS OF LOW-INCOME COMMUNITIES OF COLOR

Private companies have always engaged outside organizations to provide training services for and with them. In any city in the country, there are dozens, even hundreds, of private and not-for-profit vendors providing all manner of training services on contract to businesses, either directly or through a federal or state-subsidized program. The major federal government programs remain the Job Training Partnership Act, the welfare-to-work activities mandated by the Job Opportunities and Basic Skills program under the 1988 Family Support Act, the Carl D. Perkins Vocational and Applied Technology Act of 1990, and the 1994 School to Work Opportunities Act, aimed at integrating school-based and workplace-connected ("concurrent" or "contextual") learning. There are also many state-initiated programs that entail cooperation with private industry. All this legislation is currently being reconsidered in ongoing negotiations between Congress and the White House.

Across all these many activities, much hope is being placed in the emergence of the 2-year community colleges as crucial sources of customized training for firms and as important bridges between the streets, the high schools, the employers, and the 4-year colleges and universities. The performance of the community colleges has inevitably been uneven. Moreover, there is a lingering tension between their "vocational" training mission and those professional educators who emphasize the role of these institutions as providing remedial or transitional education for students on their way to "regular" 4-year colleges (Harrison, Weiss, & Gant, 1995, pp. 28-34). Elijah Anderson's (1990) long-standing concern about the negative consequences of a mismatch between the races of the mostly white instructors and the young African American trainees in the job training programs of the 1960s and 1970s (especially the difficulty of developing sufficient mutual trust to support effective mentoring) applies to the community colleges as well.

Notwithstanding these concerns, there are a growing number of well-documented cases of community colleges working closely with companies.[4] This has not gone unnoticed by CBOs, which are making efforts in several places to partner more closely with these key institutions that mediate between citizens and employers. Later in the book, we will present examples from our own current research of black, Hispanic, and white/Anglo CBOs involved in workforce development networks that are erecting lively and promising partnerships with local community colleges. At the same time, a growing number of companies are providing advice, instructors, and equipment to the community colleges, jointly developing curricula, and becoming acquainted with students and trainees, through instruction, summer, and "co-op" jobs, long before the latter complete their courses. In some places, 4-year colleges and universities are also making themselves key players in these citizen-neighborhood-college-employer networks. Examples offered by one of our case writers, University of Illinois at Chicago urban planner Joan Fitzgerald (1995, p. 34), are the relationships between the West Philadelphia Improvement Corporation and the University of Pennsylvania and webs linking the Universities of Kentucky and Alabama to other players within their regions.

THE TIME IS RIGHT FOR NEW APPROACHES

It cannot be stressed enough that even the most effective, most well-targeted workforce development programs will not, by themselves, solve the problem of insufficient job creation at what sociologist Lee

Rainwater called "get along" income levels. It has taken mainstream economists a long time to acknowledge the existence, let alone the centrality, of dysfunction on the demand side of the labor market. Policies and programs that fail to address the need for economic development—within inner-city neighborhoods, throughout the city, and across the whole metropolitan region—are destined to leave people with dashed hopes and continued frustrations. In the context of comprehensive economic development activities, however, the supply side of the labor market—education, training, and job search—comes into its own.

To be sure, mainstream workforce development and employment training programs have never had good press—at least partly because companies and the public have never been fully committed to them. Besides the fact that the design and management of these programs provided for the "empowerment" of a quite substantial cadre of now middle-class professionals of color at all levels of government and in the private and foundation sectors, it is remarkable how little has changed since the early 1960s. With only slight exaggeration (and the occasional error of omission), veteran ET expert and administrator Garth Mangum (1995) summed up 33 years of experience with federal workforce development policy in a lecture to congressional staffers as follows:

> Benefits have [generally] exceeded costs for adult programs, but always by narrow margins. . . . No major youth program except Job Corps has ever had a favorable ratio of benefits to costs. . . . Program proliferation has always been an irritant, interfering more [however] with orderly administration than with program outcomes. . . . The relative roles of federal, state, and local governments have constantly fluctuated. . . . The mix between preventative preparation and second chance remediation has always been an issue, as has been an antipoverty versus mainstream [targeted vs universalistic] emphasis. . . . The same can be said for welfare reform. Employed self-sufficiency for AFDC parents has long been a lodestone. Few have doubted that it was obtainable at the price of a combination of remedial education and skill training, child care and health care, subsidized private and public service employment and probable continued income supplementation for many. But so far no one has been prepared to pay the price. (pp. 4-6)

Perhaps the mainstream programs can be fixed. Perhaps the currently bleak political climate in Washington will not, after all, undermine the successful working out of the bugs in the latest ET reform movement, School to Work, with its dreams of concurrent learning (academic education plus real-world work experience), "one-stop shopping," and national educational standards.

In the meantime, there is surely no reason to hold back on learning more about, and promoting the fortunes of, more unconventional approaches to workforce development, especially those that centrally involve low-income CBOs and activists of color. The emergence of one category of such approaches—those built on interorganizational, collaborative networks—is discussed in Chapter 4.

NOTES

1. Bishop (1994a, p. 24) adds: "American employers appear to devote less time and resources to the training of entry-level blue-collar, clerical, and service employees than employers in Germany and Japan."

2. American employers do contribute part of the social cost of general training by forgoing the extra profits associated with putting up with generally less productive new workers. They are also partially recompensed, however, by being allowed—by social custom and by union contracts—to pay lower wage rates to apprentices and other novices.

3. The Bureau of Labor Statistics confirmed this well-known relationship in a recent survey of formal training programs in the private sector (see Bureau of National Affairs, 1994, p. B-19). Of course, the proportion of private businesses that are small is much greater in Japan and Germany than in the United States (Harrison, 1994, Chapter 2). Japanese and German firms, however, still manage to invest considerably more in worker training than do their American counterparts. Culture, social norms, and national public policy really do make a difference.

4. Evaluation reports and metastudies that are especially sensitive to the place of CBOs in the community college-private employer nexus include Fitzgerald (1995), Fitzgerald and Jenkins (1997), Rosenfeld (1995b), and Rosenfeld and Kingslow (1995).

WHY COMMUNITY-BASED ORGANIZATIONS ENGAGE IN TRAINING AND WORKFORCE DEVELOPMENT AND FORM OR ENTER INTERORGANIZATIONAL NETWORKS TO HELP THEM DO IT BETTER

Conventional treatments of work and labor markets place them on a flat, homogeneous terrain . . . marked only by intersections between neat curves of supply and demand. We have looked closely at the terrain and found it pitted, riven, and undulating, full of inequality, segmentation, segregation, conflict, and coercion.

—Tilly and Tilly (1994, p. 307)

Individual poor persons of color from neighborhoods thought of by employers as the "inner city" have particular trouble finding and affording quality training or getting good jobs, even if they do have training and at least some education. Why aren't skills, the acquisition of credentials, and the display of "appropriate" attitudes sufficient?

There are two conventional explanations, upon which most policy has rested for years. One explanation suggests that employers directly discriminate by race, gender, or class (Kirschenman & Neckerman, 1991). A second, more subtle, explanation is that both employers and job seekers have insufficient information about one another, and that, on average, employers do the best they can by screening in (or out) entire classes of people (Holzer, 1996). The appropriate policy remedies that have been proposed over the years include targeting government subsidies for education and training to the urban poor, improving the collection and dissemination of labor market information so that individual decisions will be less "imperfect" (e.g., by improving the Federal Job—formerly called the Employment—Service), subsidizing employers to compensate them for hiring and training people they would otherwise prefer not to hire, or making individual freedom from employment discrimination a right of citizenship such that discrimination by schools, employers, and other labor market actors becomes illegal and punishable.

Recently, a third explanation has come to dominate discussions of this subject. Perhaps the education and training being provided to low-income persons, generally, and to people of color from the inner city, especially, are out of date. Perhaps the new technologies that companies (and agencies, hospitals, and even fast-food restaurants) are introducing call for a higher degree of "skill," by which is usually meant cognitive ability: the capacity to learn new tasks and to make judgments on the job. As we suggested earlier (see note 3 in Chapter 2), the conceptual clarity and the scientific evidence in support of this supposed "skill-bias" to technological change are surprisingly weak. The introduction by managers of new technology to an individual work site is, by intention, initially labor-"saving." Whether the net result for a city (let alone a country) as a whole is additional jobs, elimination of jobs, or more or less neutral, and what these changes imply for employers' demand for skills, depends on how the productivity gains from introducing the new technology are deployed. That depends on social and political decisions about economic policy, not on the cognitive abilities of individuals per se.

The United States has undoubtedly made progress in investing in the skills and capacities of individual workers and in increasing the efficiency with which labor market institutions "match" supply and demand—for example, through the introduction of computerized information systems called "job banks." Moreover, if antidiscrimination (or at least affirmative action) laws and regulations had not been working to some degree, there probably would not be such a huge current outcry against them.

Despite several decades of individualistic or rights-based approaches to addressing urban poverty through conventional workforce development policies and affirmative action, however, conditions for those living in the areas with the highest concentrations of the poor are widely thought to have been at best only marginally improved (Kasarda, 1995). Moreover, the fraction of the population living in such places is actually rising. Ronald Mincy and Susan Weiner (1993) have shown that, between the 1970 and 1990 population censuses, the number of people living in census tracts in which at least 40% of the population were poor more than doubled from 3.8 million to 10.4 million.

We conclude that purely individual or rights-based treatments are unlikely to be sufficient to make a major dent in poverty—certainly not through conventional labor market interventions per se. Prospective trainees and workers, those displaced from farm or from older industrialized employment and needing a second chance, and long-term welfare recipients wanting (or, under the "reform" movement, being forced) to look for a job will not succeed if they only undergo some training process, get stamped on the forehead as "certified," and get sent out into the street to "make it." Certainly, we must continue to invest in skills, the capacity to learn, and attitude formation that is more appealing to the gatekeepers who control the jobs. By themselves, however, these approaches will never be enough.

The reason for this is that the conventional underlying models and beliefs about how labor markets work, especially for poor persons of color, are inadequate, if not plain wrong. For these populations especially, workers are not hired through what economists call "queues," in which workers are arranged and considered as though they were lined up, one by one, in order of their potential contribution to the profitability of the firm (or achievement of the mission of the nonprofit organization or public agency). Rather, they are hired through intersecting social and business networks of various kinds. This is not a new discovery for social scientists; we have been alerted to it at least since the researches of economists Albert Reiss and George Schultz in the 1960s (Rees, 1966; Rees & Schultz, 1970) and of sociologist Mark Granovetter in the 1970s. Until recently, however, it remained a much understudied subject, lying outside the apprehension of formal economic analysis. Even now, economic models studiously employ what has been called "methodological individualism," placing the problem strictly within the familiar utilitarian framework of individual choice under constraint.[1]

The implications of this view are profound. Not only does theorizing about social networks offer a radically different way of understanding how labor markets work in reality but also it calls for entirely different policy approaches to the problem. These approaches emphasize acting on the social structures through which people are processed rather than focusing centrally on the "disabilities" of the poor.

NETWORKS, JOB SEARCH, AND HIRING[2]

Quite apart from the difficulty and expense of physically getting back and forth between home and the job site, we now know that job seekers from low-income neighborhoods may not be able to even find employers with openings—to get onto their waiting lines—because the social networks to which these workers and their prospective employers belong fail to intersect. Also, the workers already hired by an employer do not "connect" to low-income communities of color; thus, conventional "word-of-mouth" channels for recruiting bypass the inner city altogether. The significance of such disconnects is that they leave employers and other gatekeepers with no alternative but to rely on their perceptions or beliefs about inner-city workers as a class. If those perceptions are generally negative, employers will not be open to even considering hiring them (Kirschenman & Neckerman, 1991).

Network theorizing approaches the problem from several perspectives. From that of workers, being well-connected means having a dense mix of relationships with others who are already employed (or own businesses or know those who do). Such connections can be characterized by strong or weak ties or both. Strong ties are those to close friends, relatives, or members of the same race, ethnicity, or clique. Weak ties are more voluntarily associational, referring, for example, to relationships with former schoolmates or comrades in some political or civic activity.

At least two hypotheses have been put forth suggesting that weak ties are likely to lead to greater job finding success than strong ones. Granovetter originally hypothesized that weak ties were more likely to generate a greater volume and diversity of job information so that the pure chance of at least one of those leads working out is greater. In the early 1980s, Nan Lin (Lin, 1982; Lin, Ensel, & Vaughn, 1981) suggested that the distribution (not just the sheer number) of potential job offers associated with weak ties may be superior to those learned about through strong ties. Thus, for example, as we noted earlier, young black men might in fact be very "well connected" to networks of friends and (extended) family members who already work, but the kinds of jobs likely to be obtained through these connections might not be particularly attractive. The lesson for work with community-based organization (CBO) networks is clear: Connection, by itself, is not enough. It depends on sources of information and employers to which one is connected.

From the perspective of employers, it is well-known that a substantial fraction of especially entry-level hires are made through referrals from existing employees. This is true in part because it is inexpensive. There is more involved, however, than merely cost. Tilly and Tilly (1994, p. 301) point out that recruitment networks "facilitate the creation of patron-client chains . . . and guarantee some accountability of suppliers for the quality of workers supplied" (recall our earlier observation that companies seem to be increasingly using temp agencies and other external contractors to screen prospective employees). Granovetter (1985) argues that hiring through referrals is also reliable. The reason is that trust decays with the length of the chain of contacts (that is why the impersonal computer printouts from job banks are never as persuasive to employers as information systems specialists imagine). Trust is also widely assumed to be reinforced by repeat contracting, that is, reengaging others in network transactions repeatedly.

It follows that the most helpful intermediaries—training institutes, teachers, trade associations, and community- and citywide advocates— are those that are not too relationally distant from the actual employers with the job openings. Later, we will argue that their relational proximity to and continued (not just one-time) engagement with many of the companies with which they work are the most important reasons for the universally acknowledged job training and placement success of the Center for Employment Training (CET).

Another property of networks that is germane to our work is that of individual or organizational centrality. A person, group, or institution is more centrally located within its network(s) to the extent that the most (or the most important) information passes through it. The reputation of a centrally positioned organization matters more (e.g., in making job referrals) than that of others. Power is very much about centrality. In our current research project, we have used the concept of network centrality as the principal way of organizing our case studies of regionally engaged workforce development networks involving CBOs.

Researchers in this field also ask what kinds of organizational structures internal to the firm, agency, association, or CBO are more and less supportive of external collaborative networking. How well do various internal arrangements—whether informal, top-down, or those in which the agency, firm, or community group is best seen as a network of treaties among interdependent but separately powerful departments or divisions—adjust to "ecological" (outside and environmental) crises faced by the organization as a whole? A salient example is the pressure that the need for closer collaboration between community colleges and companies is placing on the internal "equilibrium" that has developed over the years between those oriented toward preparing students to move on to 4-year colleges and the more traditional "voc ed" staff.

COMMUNITY DEVELOPMENT CORPORATIONS AND OTHER CBOs AS SOCIAL AGENTS WITHIN REGIONAL SYSTEMS OF INTERSECTING NETWORKS

It is becoming increasingly clear that there is practically no way that low-income, already socially ostracized individuals—no matter how highly motivated—can single-handedly reconstruct and negotiate a city's map of social and business connections. Thus, effective training and job placement must be mediated by collective institutions if they are to be fully effective. Individual seekers of skills and jobs must be supported by the greater economic and political power of *agents*: organizations that can break paths, open doors, insist on quality services, and negotiate collectively with employers and governments.

For the residents of low-income areas, rural as well as urban community development corporations (CDCs) and other CBOs can and sometimes do attempt to fill the role of collective agent for individual job seekers. They work with the schools, community colleges, and social workers. They are uniquely suited to provide informal recruiting and follow-up counseling because they know the trainees (or their relatives and friends). A few have become sufficiently large to be in a position to employ significant numbers of local residents in their own operations (Newark's New Community Corporation [NCC] currently has approximately 1,400 persons on its payroll, almost all of them from the black Central Ward). The most experienced of the CBOs have learned to use their political presence or organizing base as a fulcrum with which to leverage area companies and government offices to open up training and jobs to their constituents or members (as Communities Organized for Public Service [COPS] quite explicitly does in San Antonio).

The mounting of new programs, additions to in-house capacity, and flexing of political muscle by individual CBOs is still not enough—as both NCC and COPS's Project QUEST have shown. Thus, during the past 10 years, these and other community-based groups have begun to seek out partnerships, collaborations, and "strategic alliances" with other CBOs, with schools and colleges, and with private companies located within their neighborhoods, across the city, in the suburbs, and even, in some cases, across state and regional borders. In short, CBOs have increasingly entered, or created, interorganizational and boundary-spanning networks. They have done so in many fields, of which the employment training area is only one (but the one on which we focus in this book).

The question is, why? The answers do not lie within the relatively narrow domain of theories about the imperfect operation of urban labor markets. Rather, the motivations for CBOs turning to structured networking, whether informal or quite explicitly contractual, are much deeper.

From our earlier exploratory fieldwork, and from the 10 in-depth case studies that make up the heart of this book, we have learned that CBOs tend to turn to interorganizational networks in the following situations:

◆ A project, or class of projects, is too risky for any one organization to take on alone.

◆ No single organization has the internal capacity—staff, real estate, equipment, and managerial depth—to get the job done, whereas the system or network might.

◆ Key information that any one CBO needs is lodged within some other organization and cannot be easily purchased or otherwise acquired (such information is "impacted").

◆ For one organization to "do business" inside someone else's area or market ("turf"), it may be asked by representatives from that area (or from funders) to join forces with a local partner to ensure that local learning takes place.

◆ A single group is not sufficiently large, wealthy, influential, or powerful to attract a diverse pool of vendors of relevant services. By joining forces with others, the network as a whole aggregates its demands for those services, thereby attracting a more diverse mix of prospective suppliers.

◆ Gaining legitimacy in the minds of some other group of players—for example, companies with jobs—requires turning that group into "stakeholders," whose sense of "ownership" of the project is crucial for the CBO to achieve its goals.

Of course, CBOs are not always quite so strategic when they turn to networking. Sometimes, they stumble into these collaborations, as when informal, even casual relationships that have proved beneficial evolve into more structured partnerships. In other instances, smart leadership instinctively seizes unexpected opportunities to "do deals," which then develop a life of their own. Perhaps key funders have exhorted them to "collaborate," and it turns out to have been a good idea. Our cases contain many examples of all these points of entry and more.

The at least partial surrender of autonomy that goes along with entering into structured interorganizational networks makes them risky ventures, however, especially for generally fragile and vulnerable CBOs, particularly at the beginning. For those networks that have proven to be mutually beneficial over a period of time, breaking off the relationship, however sensible on its merits, is often difficult and even dangerous. The point is that, in the absence of one or more of the underlying structural reasons for a CBO to engage in boundary-spanning networks, the more opportunistic (as distinct from strategic) initiatives are not likely to be very durable.[3]

ASSESSING CBO WORKFORCE DEVELOPMENT NETWORKS

Has the growth of formal and informal, implicitly and explicitly contractual networking among organizations, and across geographic borders, significantly enhanced the ability of CBOs to win higher-quality training, job placements, and promotions (or at least long-term retention) for their constituents? What lessons can the CBOs, and also the private sector, governments, and funders, learn from both the success stories and the failures? What prior beliefs, assumptions, and theories about how networks operate does the record cause us to reconsider? Most of all, to what extent, or in what circumstances, does "messing with outsiders" (as one CDC staffer stated) in the interest of facilitating employment training objectives compromise the fundamental goal of social and economic development of the community? These are among the questions to which activists, community developers, program evaluators, and funders will surely want answers.

Unfortunately, it is still premature to offer rigorous answers to most of these questions. It is not so much a matter of there not being sufficient experience; some CBOs (such as San Jose's CET) have been managing networked relationships for approximately 30 years. Others, such as Chicago's Bethel New Life, Inc., started only recently but have rapidly grown many new, easily observable cross-organizational connections, some of which are working marvelously, whereas others have quickly collapsed. The problem is that program operators and evaluators have seldom thought about the job training process in quite this way and so have practically never asked the right questions when writing their reports or collecting their data.

In this book, we use the network concepts of strength of connection, weak and strong ties, trust, reputation, length of relational chains, centrality, and power "impressionistically" to help us give shape to our very rich (indeed, sometimes overwhelming) case study material. Frankly, we have been feeling our way along. In subsequent work, we intend to test formal hypotheses about the impacts of ties, trust, length of chains, and so on on CBO "success," which is measured by survival, the ability to build on one set of collaborative ties to develop new ones, the evolution of organizational reputation, and, of course, the incidence of placement and retention of client workers into "good" jobs. We will also make the networks themselves the object of inquiry and develop measures not just of how collaboration influences individual member organizations but also how it shapes the evolution of the network itself.

For that type of research, it will be necessary to undertake very different kinds of surveys of organizations than those that are typically developed by specialists in employment training and workforce development. Similarly, evaluations of the relative success or failure of different policy approaches to attempting to get low-income communities better (and robustly) connected to the regions in which they reside will need to employ a much broader mix of methodologies than those conventionally used in this field, featuring much greater attention to combining the quantitative with the qualitative, process with outcome, and focusing on the institutional as well as on the individual.[4]

NOTES

1. The best of this work, in which information matters, outcomes are uncertain, and social relationships constitute (highly stylized) parameters conditioning individual optimization, includes Ackerlof (1970), Dickens (1996), Holzer (1987), Mortenson (1986), and especially Montgomery (1991, 1992).

2. This section draws importantly on the pioneering work of Mark Granovetter, now a professor at Stanford University. See Granovetter (1985, 1994), Farkas and England (1994), Granovetter and Tilly (1988), Tilly and Tilly (1994), Powell and Smith-Doerr (1994), and Uzzi (in press).

3. Remarkably, much of this theory about the causes and implications of interorganizational networking applies equally well to the increasingly collaborative behavior of private corporations, both within and among countries. There is now a truly prodigious literature on interfirm networks, strategic alliances, *keiretsu*-like consortia, and industrial

districts: See Gerlach (1992), Harrison (1994), Kanter (1995), and Rosenfeld (1995a). It is beginning to appear that a new principle of both business and nonbusiness organization is emerging in the brave new world of heightened competition, accelerating movements of capital, information, and technological change.

4. For an indication of what more eclectic, practical research and evaluation might look like, see Bartik and Bingham (1995), Bartik (1995), Connell, Kubisch, Schorr, and Weiss (1995), and Hollister and Hill (1995). On the state of the art in conducting formal outcomes-based program evaluations, see Bell, Orr, Blomquist, and Cain (1995).

THE CASE STUDIES
Workforce Development Networks With Individual Community-Based Organizations at Their Hubs

Not only are organizations suspended in multiple, complex, and overlapping webs of relationships, but the webs are likely to exhibit structural patterns that are invisible from the perspective of a single organization caught in the tangle. To detect overarching structures, one has to rise above the individual [organization] and analyze the system as a whole.

Barley, Freeman, and Hybels (1992, p. 205)

THREE TYPES OF INTERORGANIZATIONAL NETWORKS

In this book, we use the analytical concept of network position to organize the 10 case studies on community-based organization (CBO) involvement in workforce development (the location of the sites appears in Figure 5.1). Using the working taxonomy that initially took us into

Figure 5.1. Networking Across Boundaries: Case Study Sites

the field (and that has held up surprisingly well, with some inevitable modifications), we first distinguish hub-spoke networks in which the CBO often holds a central, initiating position.

In other cases, networks are most usefully characterized as peer-to-peer. Although we do not expect power and authority to be symmetrically distributed among all the members of this form of network, there is typically no one dominant, central actor (unless perhaps it is the administrative staff of the network itself).

Finally, we propose a class of intermediary networks in which a non-CBO—a regional educational entity, a public-private planning agency, or perhaps a large company with strong recruiting and subcontracting networks of its own—plays the central role. In this case, the questions are whether and how well CBOs in the area become connected to the intermediary and how their interrelationships evolve over time.

Figures 5.2 through 5.4 indicate stylized sketches of these three canonical forms and list the case studies that we associate with each. In five cases, the CBOs themselves constitute the initiators of (or at least major actors within) hub-spoke networks: webs in which the CBO stands at the center of the action. These cases include the San Jose-based Center for Employment Training (CET) and Project QUEST in San Antonio, both of which are now replicating their models elsewhere; New Community Corporation (NCC) in Newark and Bethel New Life,

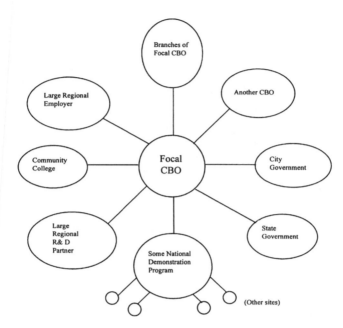

Figure 5.2. Hub-Spoke Employment Training Networks With Focal CBO as the Hub

Inc., in Chicago; and Coastal Enterprises, which operates in several small cities and rural counties in Maine.[1]

Three of the case studies—the Chicago Jobs Council, the Pittsburgh Partnership for Neighborhood Development, and the Business Outreach Centers of New York City (and, recently, Bridgeport, Connecticut)—exemplify peer-to-peer networks in which groups of CBOs work together, through a shared administrative "secretariat," to achieve the workforce development and employment training (ET) objectives that no one of the member groups can attain on its own.[2]

Finally, two of the cases place the focus on regional intermediary networks: those governed by organizations that are not explicitly "community based" but with which CBOs concerned with jobs and workforce development have partnered to better achieve their goals. One case is the Regional Alliance of Small Contractors, a spin-off of the Port Authority of New York and New Jersey that connects small and women-owned contractors to world-class construction firms and banks. The

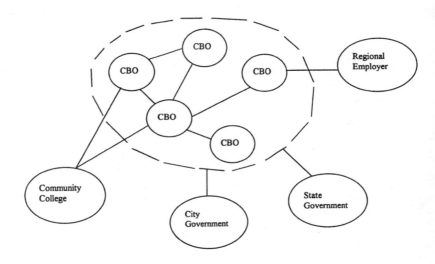

Figure 5.3. Peer-to-Peer Employment Training Networks (Webs of CBOs)

second is Lawson State, a black community college in Birmingham, Alabama, that, like many other such schools, seeks to become a source of technical assistance to small and medium-sized manufacturing firms located in its region. Modernization of the operations of these private businesses increases the chances of providing additional training and job opportunities to the families and students who make up Lawson State's constituency.[3]

These three categories—hub-spoke, peer-to-peer, and intermediary network types—are not fixed in stone. Indeed, one of the more exciting aspects of the research lies precisely in our discovery of the extent to which the most dynamic of these groups are continually trying new things, developing new partnerships, and being invited into other existing networks. What all the different types of networks have in common is an approach that emphasizes forging stronger, more effective, and long-lasting linkages to the world "outside" the neighborhoods but in the interest of further developing the social and economic

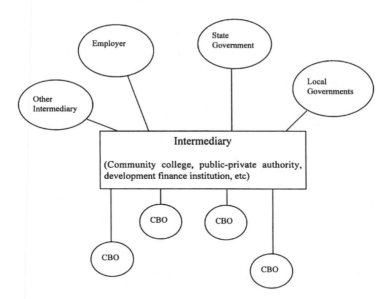

Figure 5.4. Intermediary Employment Training Networks With Intermediaries as the Hub

capabilities and well-being of the residents of those neighborhoods. Our studies emphasize the special role of such city- and regionwide partners as an area's community colleges, the more far-sighted private industry councils, port authorities, and large, influential private employers—especially the big commercial banks that have already demonstrated their understanding of the complexities of community development by taking a proactive role in relation to Community Reinvestment Act mortgage "green-lining." Some of the same banks are now assisting community development corporations (CDCs) in connecting with private firms to engage in joint economic development (and training) ventures, both inside and outside inner-city neighborhoods.

In this section, we draw on our case material from the studies of CET, QUEST, NCC, Bethel New Life, and one rural CDC, Maine's Coastal Enterprises, to offer findings on the origins, evolution, and (in a very preliminary way) effectiveness of workforce development networks with individual CBOs at their hubs.

SAN JOSE'S CENTER FOR EMPLOYMENT TRAINING

University of Massachusetts-Boston economist Edwin Melendez's
(1996) case study on the CET begins as follows:

> Based in San Jose, California, and operating (until recently) entirely
> in the western and southwestern United States [first as the Opportu-
> nities Industrialization Center of Santa Clara County (OIC-SCC), then
> as the California Expansion of the Center for Employment Training
> (CET), and then as the CET Federation], CET is one of the few training
> programs in the nation that rigorous research has demonstrated have
> a long-term impact on participants' earnings. It is the only short-term
> (6-month) classroom training model that can claim to work. CET
> trainees are placed into jobs at higher rates than [those from] other
> programs in the same cities, stay in those jobs longer than other
> trainees, and enjoy substantial gains in earnings over time. CET has
> achieved these remarkable results while serving the most disadvan-
> taged populations. Typically, trainees are dislocated farmworkers,
> mothers on public assistance, out-of-school youth, past criminal
> offenders, and individuals with limited English proficiency. For those
> who are concerned about the employability of the poor, CET offers
> hope that public and civic intervention in labor markets can make a
> difference. (p. 1)

At least prior to the replication phase, which began in 1992, CET
has continually impressed even the most hard-nosed ET program
evaluators with its solid record of success (Tables 5.1 and 5.2).
Reports conducted by or for Mathematica Policy Research, Inc., the
Rockefeller Foundation, the Manpower Development Research
Corporation, the U.S. Economic Development Administration, and
the government of Santa Clara County in California have, during a
20-year period, consistently shown CET to be highly successful, as
well as cost-effective, in placing trainees with world-class companies,
raising their earnings relative to randomly assigned control groups,
and enhancing the reputation of the organization itself within the
employer community.[4]

In 1992, under the auspices of the U.S. Department of Labor,
CET began its current replication phase. Fourteen projects were
attempted in cities along the east coast and in Chicago. Some of

Table 5.1 Evaluations of CET by Funding Agencies

Date	March-August 1971	August 1974
Evaluator	Development Associates	Office of Program Planning and Evaluation
Funding agency	EDA U.S. Department of Commerce	Santa Clara County
Universe	11 EDA-funded projects of 26 still existing centers; as of July 1970, EDA had funded 64 employment and training projects	Santa Clara County training programs
Targeted population	Disadvantaged adults; 10,657 trainees	Public assistance recipients; 400 trainees
Method	Selection of representative sample of sites: eight public work skill centers and three technical assistance training centers; analysis based on quantitative data questionnaire and site visits	WIN staff assessment and comparison to other contractors
Main findings	OIC-SCC ranked first among all projects in mean income change ($4,155 vs. $1,815 average) and in placement for the general adult population (76 vs 27.4% average); OIC-SCC also tied for first in placement of the unemployed (100 vs. 26.7% average)	OIC-SCC exceeded contracted minimum public assistance recipients training of 300 (400 trainees) and 175 placements (240 placed in jobs); East San Jose's office had a 94% placement rate of targeted population in comparison to 69% for central office placement rate

these attempts failed immediately; others have evolved in interesting ways that have changed approaches to workforce development both among the replication site administrators (private industry councils and CBOs) and within the staff of CET itself. There are instructive lessons to be learned, even from the failed efforts, about organizational learning and network governance. Currently, a formal evaluation of the CET replication phase is under way that is managed by Mathematica (Table 5.3). The number of replication sites had expanded considerably by 1996.[5]

Table 5.2 Evaluations of CET Demonstration Projects

Date	Demonstration Program from 1982 to 1988	Demonstration Program from 1985 to 1988
Evaluator	Mathematica Policy Research, Inc.	Manpower Demonstration Research Corporation
Funding agency	Four CBOs in Atlanta, Providence, San Jose, and Washington, D.C.	13 JOBSTART sites throughout the nation, including CBOs, Job Corps Centers, and so on
Targeted population	Minority-female single parent; 3,965 participants	17- to 21-year-old school dropouts; 1,941 sample
Method	Random assignment to experimental or control group; program included remedial education, job skills training, placement assistance, and child care; Control group was not eligible for services at participation experimental sites but could seek similar services elsewhere	Random assignment to program participation or control groups; participants received combination of skills training, basic education, support services, and job placement; control group could not participate in JOBSTART activities but could seek similar services elsewhere
Main findings	In a follow-up 30 months after program participation, CET raised monthly earnings by $100 and $2,000 over the period; program raised earnings particularly for those with 12 years or more of schooling and for those out of work the previous year and receiving welfare; researchers suggest that the "integrated learning" design could be responsible for program impact	CET earnings impacts (of $6,700 for 4-year period) were the largest of any other site; program design did not seem responsible for program impact—there was no significant difference in program outcomes between concurrent or "integrated" education training programs and "sequential" ones

For the purposes of the roughest of comparisons, the Office of the Chief Economist of the U.S. Department of Labor concluded that CET netted an increase in the early 1990s of approximately 33% in the annual earnings of out-of-school youth beyond what might have been expected, whereas the Job Corps' comparable rate of return was approximately 15%.[6] Similarly, among the four sites evaluated in the Minority Female Single Parent Demonstration

Table 5.3 Evaluation of CET Replication

Date	Winter and spring of 1994
Evaluator	Mathematica Policy Research
Agency	U.S. Department of Labor
Universe	10 CET replication sites including CBO-, JTPA-, and CET-operated sites
Targeted population	Organizations replicating the CET model
Method	Site visits and interviews
Main findings	Overall, sites fell short of demonstrating the replication of the CET model; results suggest that this seems to have been a premature evaluation of replication efforts but lay out a conceptual framework for future evaluations

Project of the Rockefeller Foundation, 30 months after beginning a job, CET's trainees were the only ones to show positive results, with annual increments to income of approximately $2,000 and slightly increased expected employability. Also, in the national JOBSTART evaluation conducted by MDRC, CET's average impact on net earnings fully 4 years after placement averaged an additional $6,700 of earnings—far greater than the performance of 13 other programs for training criminal offenders and high school dropouts (U.S. Department of Labor, 1995).

A program that so successfully deals with new immigrants, single mothers with small children, youth, ex-offenders, and former farmworkers, often coupling the training with both classes and on-the-job English as a Second Language, stands out amid the myriad of more specialized attempts during the past quarter century to significantly affect the working lives of poor people. Why does it work—or at least why has CET been so successful prior to its replication phase (regarding this, the jury is still out)? On the basis of 2 years of direct observation, Melendez's (1996) conclusion is that CET has found a way to successfully motivate its trainees to seriously develop job-specific skills and has made itself an integral part of the recruiting networks of employers. In other words, CET works on both the supply and the demand sides of the labor market, and its approach to the latter explicitly recognizes the importance of inserting itself into precisely those networks that companies already value. Moreover, CET grew out of the social unrest and political organizing of West Coast Hispanic farmworkers in the 1960s, and its legitimacy within the now widely distributed Hispanic population—which, Melendez believes, helps to explain its ability to especially motivate Mexican-American trainees—continues to derive in part from its self-definition as part social movement.

What do companies receive from working so closely over so many years with a CBO such as the CET? Without a doubt, they value the fact that CET provides screening, reliability, and follow-up counseling and supplies relatively high-quality but low-wage (although generally above poverty level) labor. As shall be shown, however, there appears to be more to it than that.

Elements of the CET "Model"

Figure 5.5 displays the principal elements of what most outside observers, evaluators, and the CET staff itself have come to refer to as the "CET model." Contextual along with classroom training—what educators call "concurrent learning"—has proven to be especially effective at motivating populations that have had mostly unpleasant past experiences with regular classroom settings. The open entry-open exit approach, which explicitly rejects "creaming" in the interest of showing funders and evaluators "good numbers," allows people, regardless of life course and capabilities, to proceed at their own pace and considers "graduation" to occur when the trainees are successfully placed into regular jobs and not before. Trainees move from one level of skill development to another based on demonstrated competency, as judged by supervisors who, more often than not, have been recruited from the pool of companies that will eventually hire the trainees. An observer is invariably struck by the extraordinarily tight degree to which these elements are integrated in a typical day in a typical CET facility.

One particular aspect of open entry-open exit is worth noting. This approach effectively staggers graduation. That is, at any moment, certainly within a short period of time, a trainee-in-process

Figure 5.5. The CET System

sees people around him or her actually leaving to go into a job. It is not necessary to wait until a formal date in the future to see this occurring. The potential for motivation would be hard to overstate.

Although CET's combination of contextual learning, open entry/open-exit, and competency-based training is striking, none of these elements is unique. For example, concurrent learning has become the norm among best-practice programs and in the theories of professional educators. Moreover, on the surface at least, CET is really only practicing mediation of the employment relation, serving as a broker to link up particular populations to companies with job vacancies. Professional head-hunting firms, high school and college guidance counselors and placement officers, and even the U.S. Job (formerly Employment) Service have been doing this for many years. Although targeting these services to very poor and marginalized populations certainly is relatively rare, in the last analysis we do not think that the much praised CET model as such explains its unusual success as an organization.

Behind the Technical Evaluation: The Institutionalization of Training as Work

We trace CET's extraordinary success to two qualitative characteristics of the organization. First, as already suggested, CET more than almost any other training program in modern American history, has profoundly institutionalized the process of interfacing with the already trusted recruiting and training networks of companies. This institutionalization of the employer ethos extends to the daily routines within CET training facilities, which resemble ordinary workplaces (complete with time cards and a paycheck rather than a credential) and are peopled by experienced instructors from the companies. Indeed, most work-related difficulties (and even "outside" personal problems) are handled by the immediate "supervisor" (the instructor). A day at a CET facility feels almost exactly like one spent in an on-the-job training session at a major corporation.

Another manifestation of this institutionalization of the "employer mentality" is the key role of the Industrial Advisory Boards (IABs) and the Technical Advisory Committees (TACs). Each site has its own IAB and TACs (although some obviously work better than others) and comprises corporate executives, human resource managers, first-line supervisors, and even engineers. The best of the IABs are highly structured, meet often, and engage (and sometimes take the lead) in curriculum development, fund-raising, and seeking or donating their own equipment. At one training session for custodial work that we attended in San Jose, IAB members gave CET staff the most thorough lecture on the proper and safe uses of workplace chemicals that we had ever witnessed. The term *stakeholding* has been much overused and abused in this field, but it is clear that companies, including IBM, Hewlett-Packard, Motorola, Lockheed,

Price-Waterhouse, Pacific Telephone and Telegraph, FMC, United Technologies, Container Corporation of America, General Electric, and Manpower, Inc., see themselves as stakeholders in CET's success and have, during the past two decades, acted accordingly.[7]

Upon opening a training center, CET staff immediately begin dialogues with human resource and other managers from the area's private firms. In forming its TACs, not only is CET assessing job opportunities and identifying appropriate occupations for training but also it is identifying receptive managers upon whom its job developers may expect to continue to call for engagement in their process. Although curriculum development involves the usual technical business of selecting textbooks and preparing materials, in CET the principal activity at this stage is enlisting (or outright hiring) instructors from the area companies and checking with engineers and others to be certain that trainees will be working with the same kinds (and generation) of equipment—computers, machine tools, laboratory equipment, and software—that companies regularly employ. Instructors are expected to have a minimum of 5 years of experience in the relevant industry; most have many more. All these relationships may be said to be dialogue intensive.

Embeddedness Within a Social Movement

The second aspect of CET's success thus far, missed entirely in the more strictly technical evaluations, has to do with empowerment and respect derived from the organization's long-standing relation to (embeddedness within) West Coast Hispanic politics and culture. Simply stated, CET derives great strength from its association with a powerful social movement. Many of the same forces and actors who created or sustained the modern farmworkers' organizing activities, and eventually the United farmworkers of America (UFW), played a role in the formation and sustenance of CET.

In his analysis of the origins and evolution of the UFW as a social movement rather than simply another labor union, Marshall Ganz (1995) begins by adopting the observation, associated with the "structuration" school of sociology, that actors, with their identities, strategies, and interests, interact constantly with resource-endowed institutions governed by both formal and informal rules and linked to other organizations and institutions by both internal and external networks.[8] During "settled times," when people more or less accept the existing structures and rules of the game or at least share the prevailing assumption that changes are really not possible, institutions strongly shape and constrain the behavior of actors.

According to Ganz's (1995) formulation, however, in "unsettled times," when the prevailing order is being strongly challenged perhaps by the onset of forces or shocks from outside as well as inside the system, strategically placed actors can profoundly change the shape and future development of institutions. Moreover, in the process of transforming preexisting institutions, actors in critical moments need to construct new narratives to make sense of the new conditions in which they find themselves and the new possibilities for change—what Michael Piore (1995) and Renato Rosalso (1989) call "bricolages." Figure 5.6 sketches these relationships.

Those actors who are least wedded to the status quo, and who belong to many intersecting social networks without being central to any one of them—the so-called "borderlands" actors—are often the most effective in leading and shaping institutional change during moments of crisis in the prevailing order. Thus, in the case of the campaigns to organize the mostly Mexican farmworkers on the West Coast during the 1960s and 1970s, for all the efforts of the mainstream American trade union movement, its Agricultural Workers' Organizing Committee (AWOC) operated largely within the system or marshaled outside actors (such as Filipino crew leaders and supervisors) who were decidedly not members of the communities that the Mexican and Chicano farmworkers themselves constituted. By contrast, the UFW drew its organizers from those communities mobilized on the basis of social and cultural symbols and institutions already trusted by those communities (e.g., the Catholic Church) and clearly constituted a borderlands actor in the sense of belonging to a wide variety of networks, including the civil rights community, student radicals, and, of course, the labor movement (Figure 5.7). In Ganz's (1995) words,

> The AWOC was led by white male trade unionists in their 50s and 60s, responsive to the organization paying their salaries, who viewed

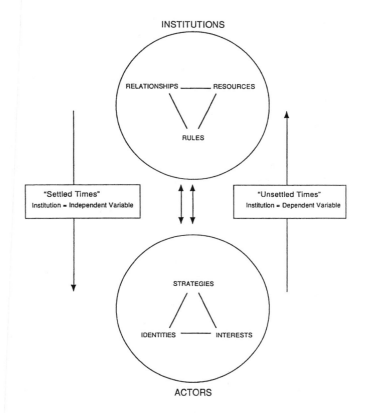

Figure 5.6. Institution-Actor Relationships
SOURCE: From Ganz, 1995

the farm labor situation through highly institutionalized lenses and whose predictable strategies were based on "taken for granted" assumptions about the situation in which they found themselves. The UFW, on the other hand, was led by a younger generation of men and women drawn from the ranks of farmworker families, trained as community [rather than "labor"] organizers, for whom organizing farmworkers had become a personal "mission," and who developed new strategies to grapple with a situation about which they had few "taken for granted" assumptions. Situated in the institutional border-lands, they constructed a strategic "bricolage" from elements at hand: the world of community organizing, efforts of activist clergy with

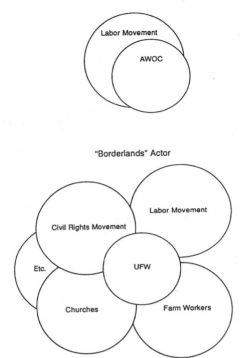

Figure 5.7. Institutionalized Actor and Borderlands Actor

whom they had become acquainted, and the cultural traditions of the people among whom they had grown up. These sharp differences in identity were further reflected in selection of organizers, worker leadership, lawyers and supporters. (pp. 18-19)

The famous Delano grape strike against the Schenley Company, which strategically incorporated the support of other unions (especially Walter Reuther's United Automobile Workers) and such friendly national political allies as New York Senator Robert F. Kennedy, offers a concrete illustration of a union organizing as a social movement. Consider, for example, the launch on March 17, 1966, of the 300-mile march from the fields of Delano, in California's agriculturally rich Central Valley, to the state capitol of Sacramento. Ganz (1995) recalls,

The march was conducted during Lent and timed to arrive in Sacramento on April 10, Easter Sunday. A farmworker led the march carrying a banner of Our Lady of Guadalupe, the patron saint of Mexico's poor, portraits of campesino leader Emiliano Zapata, and banners proclaiming "peregrinacion, penitencia, revolucion" ("pilgrimage, penance, revolution"). . . . Although invited to participate, AWOC leaders declined, declaring it was "a labor union, not a social movement." The march attracted wide public attention, particularly after Delano police attempted to block it, and, on April 3, after mediation by the Los Angeles representative of the Hotel and Restaurant (bartenders) union, Schenley recognized the UFW and signed a contract covering its grape growers in Delano. The 82 original farmworker marchers reached Sacramento accompanied by 10,000 farmworkers and supporters. (pp. 21-22)

The Evolution of Centro de Estudios Para Trabajo as an Organization

CET evolved out of these very same struggles. Founded in 1967 as the West Coast branch of OIC, the organization was led from the very beginning by a charismatic Franciscan priest, political activist, and PhD in sociology Father Anthony Soto and by Russ Tershy, a veteran of the Christian Workers Movement and the Peace Corps (Tershy is still executive director of CET). The first home of OIC-SCC was on the grounds of Our Lady of Guadalupe Church, a trusted community center in the Sal si puedes ("get out if you can") barrio of east San Jose. Cesar Chavez was a frequent visitor. Initial funding and in-kind assistance came from the San Francisco Archdiocese of the Catholic Church and from a number of Silicon Valley companies, which were just beginning their remarkable long wave of rapid economic expansion and much in need of well-trained but inexpensive blue-collar employees (Figure 5.8).

The ethnic and social movement connections were central to CET's breaking from the largely African American and East Coast-dominated OIC by 1976. Chicano groups throughout California and the farmworkers' movement brought pressure to bear on the Nixon administration in Washington to provide funds for training displaced immigrant and seasonal farmworkers and assisting them to make the transition to industrial and urban occupations. The first U.S. Depart-

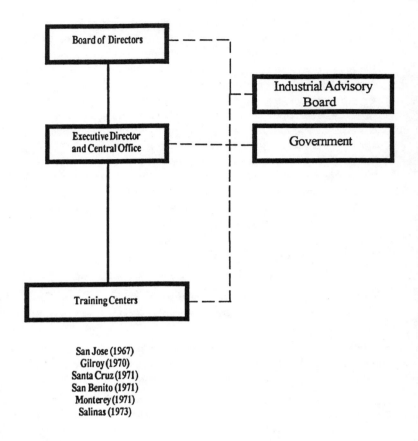

Figure 5.8. OIC-SCC (1967-1976)

ment of Labor grant to CET was made in 1973. By 1976, there were CET training centers throughout California, in Washington, and in Idaho (Figure 5.9).[9] Within 5 years, a "federation" of centers, some with their own boards of directors operating on licensing agreements with CET but most set up as (or eventually turned into) branches centrally controlled from San Jose, was sprouting throughout the West and Southwest and has continued to grow (Figure 5.10). Although it is not uncommon for such organizations to display

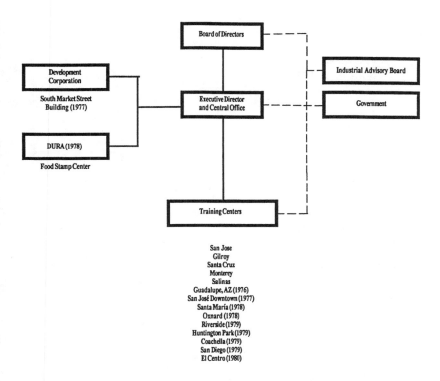

Figure 5.9. The California Expansion (1977-1980)

ethnic and racial art and historical documents on their walls, that CET's headquarters in the former Woodrow Wilson Junior High School complex in San Jose is amazingly graffiti-free and full of neighborhood residents coming and going at all times of the day speaks to the esteem with which CET is held by its community. Moreover, CET continues to this day to receive funding from the federal government for training farmworkers; therefore, that long-standing connection continues to be nourished.

The 1980s was a period of crises and occasional setbacks for CET, with the painful replacement of the more liberal and expansive Comprehensive Employment and Training Act employment training

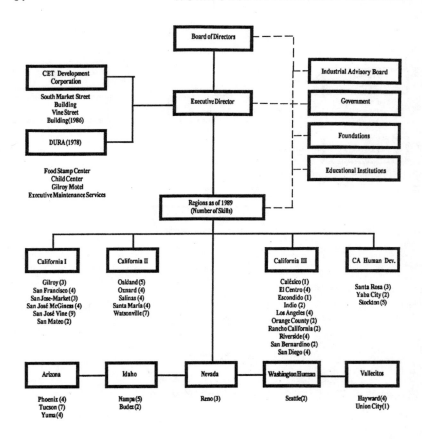

Figure 5.10. The Federation (1981-1992)

regime by the more threadbare Job Training Partnership Act (JTPA) system and with a serious and unexpected (but fortunately short-lived) recession in the high-tech sector that contained so many of CET's network partners. Compared with many other CBOs, however, CET's long-standing habit of collaborating with the private sector stood it in generally good stead with the private industry councils that were created to govern JTPA. It was during the JTPA years of the 1980s that CET expanded into the area of welfare-to-work training and became the nation's second largest vendor of legalization and English as a Second Language services for immigrants. What had previously been a rivalry with the community college sector evolved into a "cooperative competition" that brought with it California state accreditation for continuing education, giving CET trainees access to federal (U.S. Department of Education) Pell grants.

The Department of Labor Replication Phase and Its Contradictions

More than any other single achievement, it was CET's success in training and placing single mothers on welfare, as documented in the Mathematica evaluations of the female single-parent demonstration sponsored by the Rockefeller Foundation, that led to the decision by the Department of Labor (DOL) in 1992 to provide grants to JTPA agencies in Chicago and along the East Coast to work with CET to "replicate the model." A second round of replication grants was awarded in 1994 (Figure 5.11). A formal evaluation of this replication process is currently being conducted by Mathematica, whose initial "read" in 1994 identified a number of potentially serious problems.

As we were informed during one of many site visits to San Jose, many of these problems were anticipated by CET leadership and the board of directors, who have hotly (and, as is their fashion, openly if only internally) debated entering into the licensed replication phase ever since it was first proposed by the Bush and then Clinton administrations.

Although definitive results are not yet available, Melendez's (1996) case study offers several conjectures as to what may be going wrong. First, whereas the long-standing open entry-open exit approach embodied and reflected the almost entirely voluntary nature of people's participation in the training, the context for the current replication phase is welfare "reform," with its legal requirement that people take training as a condition for being able to continue to receive support for their children. Thus, classrooms at the replication sites already contain, and will increasingly contain, a mix of men and women who are there because they want to be (most of whom will

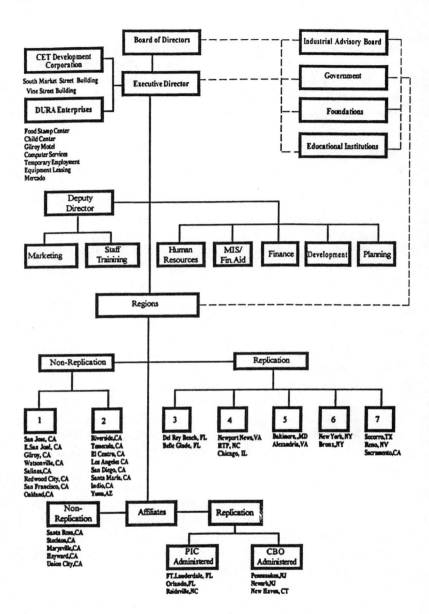

Figure 5.11. The National Replication (1992-present)

have had some previous work experience) and women, with practically no work experience, who have no choice but to be there. Melendez also observes that with a few notable exceptions, the local host group in the current replication is more linked to the private industry councils (PICs) than to indigenous community groups. With the exceptions of New Haven, Connecticut, and Pennsauken and Newark, New Jersey, where OIC (and, in Newark, the New Community Corp.) are the licensed partners, the DOL requests for proposals for replication grants have been directed at PICs that can guarantee a certain level of funding support for the project.

The first phase of the Mathematica evaluation reported that, as of 1994, links to area employers at most sites were still weak (serious IABs had yet to be established), occupations tended to be selected through formal surveys of the local labor market with "scant employer participation," trainees were typically precluded from switching skills if the first choice did not work out, complementary support services such as child care were often unavailable, and few training sites truly simulated recognizable work settings. Basic skills were still being offered in "companion" classes rather than taught concurrently (Hershey & Rosenberg, 1994).

From his recent visits to a number of replication sites, Melendez (1996) finds support for many of these observations but also some signs of change. He expresses concern that PIC-related program operators often appear to perceive CET as just one more training vendor, with its famous model being reduced to another mere collection of techniques for providing training "services." To use a theoretical construct currently in fashion, it is being *disembedded* from its holistic, cultural, and political context.

Melendez (1996) suggests, however, the collaboration with NCC in Newark is based on unique circumstances and may be developing in promising new directions (a conclusion that our case study of NCC, by Ann Griffin, strongly confirms). As will be shown when we present that case, NCC's exceptionally large scale—there are currently approximately 1,400 people employed in its own administrative, child care, transit, housing rehab, commercial development, and airport and seaport laboring enterprises—has both allowed and encouraged its staff to direct their own preexisting workforce development efforts toward staffing their own "internal labor markets." Of course, as a "player" in New Jersey-New York politics, NCC has cultivated close working relationships with many large corporations, government agencies, and the nearby community college. Still, its job training focus had been somewhat parochial. The CET collaboration seems to have impressed upon NCC's director and staff the need to relate to ("seed") the larger regional labor market outside of Newark's Central Ward, let alone NCC enterprises, alone. For its own part, CET's leadership has had to learn how to relate to a powerful, mostly African American organization that has its own power base and strong philosophy about jobs.

It appears that CET, aware of the problems in so many of the replication sites, is attempting to take greater control of the process. Moreover, it is doing so in partnership with NCC. For example, a new CET-managed "council" of replication sites (inspired by the model of the successful West Coast and southwestern CET federation?) has been meeting periodically since June 1995. The meetings have been held at the NCC training site in Newark.

CET, Network Theory, and the
Future of the National Replication

The CET offers material for reassessing all the cutting-edge theories about social and interorganizational networks about which we have written in this book. Thus, with respect to its ethnic and political constituencies, especially on the West Coast, we have found CET to be embedded within a powerful Hispanic social movement. Its legitimacy and its ability to motivate otherwise hard-to-motivate populations are almost surely associated with what Granovetter (1985) would describe as a mix of both "strong" and "weak" network ties based on family, extended community, referrals by ethnically specific word of mouth, and visibility as a contributor to local housing and community development needs of former farmworkers and low-income Hispanics, generally. The "long chains" of connection—from the farmworkers' union and social service agencies to immigration mediators to the labor pools from a wide array of towns and villages—extend throughout the American West and Southwest and into Mexico.

By contrast, CET's ties to its business "customers," especially in Silicon Valley, began in the 1970s with what Granovetter (1985) would call relatively impersonal or weak network ties, characterized by short information chains. As AnnaLee Saxenian (1994) and others have noted, the high-tech Silicon Valley companies constitute Granovetterian business groups, linked by a mix of strong and weak ties and embedded into a shared culture with a great deal of interfirm mobility by programmers, managers, and systems analysts. CET's approach has been to gradually penetrate this cluster of companies by working closely with a few firms, developing trust, and gradually transforming weak ties into strong ones—literally becoming part of the procurement and human resources systems of the valley.

How have they accomplished this? How do they seem to be pursuing the strategy in at least some of the East Coast replication sites? Initially, a few key companies are sought out and brought into the CET process. These few act as first-order contacts for the CBO with the rest of the business community. Relations with other companies then evolve from these key or core contacts. Employers' confidence is enhanced by their active engagement on CET's Industrial Advisory Boards (modeled on the even older approach of the OICs). CET presses its company partners to become actively involved in curriculum design. The companies sometimes provide instructors or equipment for use in CET training centers.

In all these respects, CET's practices on the demand side of the labor market are consistent with both Granovetter's (1985) hypothesis about the "strengths of weak ties" and Burt's (1992) powerful hypothesis that the "chasm" that a weak tie "bridges" constitutes a kind of "hole" in the social structure whose very openness creates an opportunity for entrepreneurial initiative, learning, and innovation. To at least some degree, it is plausible that other CBO workforce development networks combine these qualities of having long chains with both weak and strong ties connecting the CBO to its principal labor pool and short chains with initially weak, then (if they are successful) gradually stronger ties between the CBO and the community of employers. We think that this idea of an asymmetry between the network structures on the supply and demand sides of the labor market has not been proposed or examined heretofore.

The leadership of CET is quite self-conscious about the tensions inherent in its approach to organization. On the one hand, from its inception as a product of farmworkers' organizing struggles in California in the 1960s, CET has been a borderlands actor, embedded within a number of intersecting social and political networks but central to none. On the other hand, CET tends to govern its federation of project sites as a centralized corporation, with San Jose staff and organizers providing resources, guidance, technical assistance, and (sometimes) discipline to the local projects. This tension between being an "insider" and an "outsider" has only been heightened by its national visibility and by its quite unique role within the current replication phase.

As we indicated earlier, mainstream programs (before and since the advent of the PIC system) continue to be held in generally low esteem by most employers. By contrast, Melendez (1996) reminds us that before the research community began emphasizing the importance of networks, CET was building its West Coast strategy on the assumption that improved access to even low-end jobs in local labor markets requires workers to have connections to the recruitment networks used by employers to secure what is, in their view, a reliable labor force.

More than any other single factor, we expect the success of the CET replication phase to ultimately turn on whether the new sites can gain that confidence of and insert themselves into the trusted recruiting networks of area companies.

PROJECT QUEST AND THE COMMUNITIES ORGANIZED FOR PUBLIC SERVICE AND METRO ALLIANCE OF SAN ANTONIO

Another in a long string of devastating San Antonio factory closures, the 1990 shutdown of a Levi Strauss plant permanently displaced approximately 1,000 mostly Hispanic women. At this point, two local priests, Fathers Al Jost and Will Wauters, sought out the leadership of the Southwest Industrial Areas Foundation (IAF; an affiliate of the national IAF community organizing network, which is directed by Ernesto Cortes, Jr.) and its two local membership groups, Communities Organized for Public Service (COPS) and the newer Metro Alliance. What commenced was an extraordinary grassroots-based reexamination of the jobs and training situation in the area.

A 2-year analysis of the local labor market was undertaken with the help of professors and other labor experts at the L. B. J. School of Public Affairs at the University of Texas in Austin (where Southwest IAF is based), especially Ray Marshall (President Jimmy Carter's secretary of labor) and Bob McPherson, and involving dozens of house meetings with the mostly Mexican and Chicano members of COPS and with Metro Alliance's largely African American constituency. City leaders, then Texas governor Ann Richards, other state officials, the local community college system, the Private Industry Council, and leading local bankers were all enlisted in an effort to in effect create a new workforce development intermediary to supplement San Antonio's official JTPA system.

In her case study for our project, Dr. Rebecca Morales, University of Illinois-Chicago Associate Professor of Urban Planning and Policy and former executive director of the university's Center for Urban Economic Development, reports how Project QUEST (Quality Employment Through Skills Training) was officially launched in January 1993. For its conventionally measured achievements in job placement, significant postplacement wage gains for trainees vis-à-vis a statistically constructed comparison population, and reduced welfare dependency—recently given high marks by Massachusetts Institute of Technology's Paul Osterman in its first formal program evaluation (Osterman & Lautsch, 1996)—and for its self-conscious embedding within networks connecting the churches, area companies, and organs of state and local government, QUEST stands out as another authentic community-based network organization within the field.

QUEST as Labor Market Intermediary

Thus far in its short lifetime, QUEST has managed to package resources from a wide variety of sources, including Pell grants, JTPA and CDBG funds, and unprecedented grants from the General Fund of the city of San Antonio. In contrast to most JTPA-sponsored training programs for adults, whose clients participate for an average of 4 months (Osterman & Lautsch, 1996, p. 1), and also differing from CET's open exit-open entry "train until you get a job" model, QUEST brokers nearly 2-year-long community college-based training for somewhat better educated, often more experienced workers who have encountered chronic difficulties in moving out of poverty-level jobs. The program requires a high school diploma or a GED certificate of all applicants, although some remedial education is permitted—and made available—on the side (90% of QUEST students undergo remedial English, math, or both; Osterman & Lautsch, 1996, p. 16). In fact, Osterman and Lautsch found that 45% of the trainees already had some college.

The QUEST staff conducts occupational analyses of the local labor market, working together with personnel from St. Phillips' College, Alamo Community College, other community college personnel, and with companies in the region. Indeed, as with CET, private-sector businesspeople play a major role in identifying target occupations and in designing the training curricula. Large firms are invited to forecast sectoral demand. Initially, three broad sectors were selected for training: the health professions, business services, and environmental technologies. Partly in response to both difficulties (such as the managed care-motivated cutbacks in health-sector hiring) and opportunities (notably the enhanced demand forecasts for workers in such North Atlantic Free Trade Agreement-related occupations as truck mechanics), QUEST staff are currently broadening the menu of occupational choices.

The more than 800 trainees enrolled by Project QUEST between its inception in 1993 and the end of 1995 were provided a comprehensive, expensive package of supports, including child care, transportation assistance, medical care, tutoring, modest cash assistance for incidentals, and tuition to community college. Average cost per trainee on an annual basis (including overhead) was approximately $7,200 (Osterman & Lautsch, 1996, p. 57). Of the 629 participants no longer in training, 447 were placed into a job, joined the Armed Forces, or entered another training program (this is how JTPA officially defines a "positive termination").

Those who actually found jobs saw their hourly wage and annual earnings and average hours of work per week all rise, both absolutely and relative to the averages for comparable employees in Bexar County (Osterman & Lautsch, 1996, pp. 3-4). Gains were greatest in the health-related jobs. As with CET, gains seem evenly distributed among trainees with varying backgrounds; everyone benefits more or less equally (women tend to gain absolutely more than men but start from a much lower pretraining base). Project QUEST's average annual rate of income return to individual participants was "far greater" than the typical national payoff to an extra year of schooling, which a host of studies suggests to be in the range of 7% to 10% (Osterman & Lautsch, 1996, p. 59).

In their own interviews and research, Clark and Dawson (1995, pp. 21-22) agree that QUEST has achieved an enviable record as a job broker or intermediary within San Antonio in a remarkably short period of time. They are more skeptical of whether QUEST is also making progress toward its own explicit objective of prompting institutional reform among mainstream actors and agencies. Both Morales and Osterman and Lautsch (1996), however, remark on the positive impact QUEST has had on making the community college system more creative and responsive. Osterman is especially enthusiastic about QUEST's systems transformative impacts on the community colleges.

In the same vein, Morales discovered that, for the purposes of forecasting demand, QUEST has been able to organize stakeholder consortia of smaller companies that would otherwise have found it impossible (or distasteful) to collaborate with an overtly political organization such as COPS. Our interviews with local bankers and other businesspeople transparently revealed the hand of IAF, COPS, Metro Alliance, and QUEST in helping local elites to refashion their characterization of the area's competitive advantage from one based on "cheap labor" to an image turning on a combination of an increasingly skilled workforce and strategic access to the Mexican market.

Political Economy of Project QUEST

As with CET, with whom COPS shares a political history dating back to the era of farmworkers' organizing and the spirit of *La Raza* in the early 1970s (Rogers, 1990), perhaps the most interesting aspects of QUEST's evolution are political rather than narrowly economic. Like CET, QUEST has a strong ethnic identity, grounded mainly in the Hispanic activist Catholic Church. Both CBOs have used that political clout and social capital to get and hold the attention of mainstream actors and to build durable intersecting networking relationships. Although the particular context differs from CET, Morales also sees QUEST as "lying at the intersection of [several] stakeholder networks, and as such is embedded within them" (Figure 5.12). Much of her case study is devoted to exploring the history and evolution of these connections.

Whereas CET makes use of (but does not aim quite so overtly at) community empowerment per se, QUEST was designed from the beginning with a larger objective in mind and has built-in operating procedures for achieving it. The two Western and largely Hispanic CBO organizations differ in other ways as well. For example, whereas CET is famous for offering prospective trainees a menu of occupations from which to choose—and, in Melendez's (1996) judgment, is likely to be less successful at replication sites at which its projects lack sufficient scale to be able to do so—QUEST assigns trainees to occupations based on experience and formal aptitude and skill test results. For all its explicitly "political" character, however, QUEST is professionally managed by experienced (hired) veteran human resource specialists drawn from the U.S. Air Force—a major presence in the region since World War II (the initial director was Jack Salvadore; the current director, who had been Salvadore's number two, is Jim Lund).

Prospective trainees—"QUESTERS"—typically hear about the program through, or are explicitly recruited by, COPS or Metro Alliance

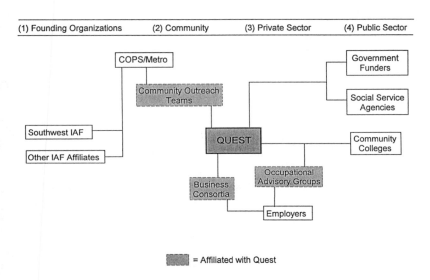

Figure 5.12. The Network Structure

neighborhood committees. They are initially interviewed by their neighbors in churches and community centers. These neighbors, along with QUEST's badly overworked, amazingly dedicated professional counselors, act as mentors, meeting with trainees at least once a week and sometimes (but not often enough, we were told) extending to postplacement follow-up. Osterman and Lautsch's (1996) surveys showed that the great majority of QUESTERS who completed the program rated the role of their counselors as "very important" to their being able to get through it.

Community residents who are members of one or the other of the organizations are empowered in the program design to "screen in" applicants who not only show a willingness to commit to a rigorous training regime but also are likely to "pay back" by helping to recruit others and by speaking about their experiences to future recruits. Taken together with the usual incidence of early dropping out, the relatively high education threshold, and the use of formal testing, it is estimated that only approximately one of five of those who originally show up for interviews eventually go into training (Osterman & Lautsch, 1996, p. 11). Indeed, this aspect of the design has led some mainstream ET program evaluators to conclude that QUEST is sufficiently guilty of creaming as to require discounting any benefit-cost measures of its performance. Osterman's response to this charge is to assess the extent of the barriers that the QUESTERS face in the local labor market, from which he draws the qualitative judgment that they would still not have done as well in the absence of this institutionalized intervention.

We concluded that the creaming charge makes sense only if the program is (mistakenly) characterized narrowly as purely a "skills training" venture. The larger objective of adding to IAF's cadres of mobilized, engaged citizens must be taken into account. In other words, training must be seen as another means to that larger end, as had earlier COPS "actions" in demanding better public sanitation, housing, public libraries, clinics, parks, and street lights or in its recent campaign to block expensive tax abatements for wealthy corporations. Osterman and Lautsch (1996, p. 1) acknowledge that "in an important sense, Project QUEST is about community building, not just job training," although their surveys suggest that, thus far, QUEST has had only a "modest impact upon participants' involvement in community affairs" (p. 5). Osterman and Lautsch also found evidence that the children of participants are more likely to pursue additional education themselves, and that welfare dependency falls beneath the rate that would have been expected for a similar population.

What Next for QUEST?

Organizers are already replicating QUEST in Dallas, Fort Worth, and in the lower Rio Grande Valley. IAF organizations in at least nine other cities (including Baltimore) are examining QUEST as a possible model for their own employment training initiatives. How unique has the San Antonio experience been? Can it be a replicable model?

Morales is hopeful but skeptical. She observes that funding is (as always) crucial and believes that municipal funding would not have been forthcoming in San Antonio without the special political relationship between COPS and then governor Richards. Most of all, she questions how many other prospective sites have the combination of intersecting supportive networks in place and the deep commitment to grassroots democracy that, above all, explains the project's legitimacy, especially in the Mexican and Chicano communities. She insists that QUEST really was designed from the ground up, using outside advisers as exactly that; it is in no sense a typical top-down "antipoverty" program. None of this is impossible elsewhere, especially where there is already a lively IAF presence. These are, she suggests, formidable prerequisites.

Meanwhile, in San Antonio itself, all is not entirely well. At the narrowly programmatic level, it is disconcerting—and costly—that "the average negative terminator had been in the program for 14.6 months prior to leaving (compared to 17.7 months for positive terminators)" (Osterman & Lautsch, 1996, p. 61). This means that almost as substantial an investment is being made in those who drop out as in those who are (according to the official definition) "successful." Moreover, managed care-induced restructuring in the health care field, plus the exhaustion of what had always been short-term prospects for substituting local trainees for Asian and other immigrants who had been obtaining local nursing and other health-sector jobs, greatly increases QUEST's need to broaden the range of occupations for which it trains. In this context, the recent truck diesel mechanic program with St. Phillips' College is most promising.

With state and federal government cutbacks, funding for QUEST is likely to become increasingly unstable and unpredictable. Osterman and Lautsch (1996) suggest several possible responses, from the provision of a broader range of services to employers (paid for by user fees) to what they call "remobilization." This last point bears further examination.

Former Texas Employment Commission official Barbara Cigainero (as quoted in Reardon, 1995, p. 7) remarked in early 1995 on the extent to which IAF's, COPS', and the Metro Alliance's direct political engagement of the private sector was a key element in how quickly QUEST was able to get up and running: "The commitments that they secured in advance from employers [so-called partnership agreements] are unparalleled. I don't know of any program in Texas with that kind of a record."

As QUEST has become successful, COPS and the Metro Alliance have increasingly disengaged from day-to-day involvement in the project. Network relationships with employers threaten to deteriorate to the point where commitments of future job openings may be harder to obtain. The political and cultural strengths and influence of IAF and its Texas affiliates have actually not been weakened by the change in electoral fortunes in the state; Governor Bush is a big supporter. Membership continues to grow. Osterman and Lautsch (1996) suggest that perhaps it is time to reinject some of that grassroots democratic political pressure back into the process—a position with which Morales would surely concur.

EXPANDING THE CAPACITY TO DO WORKFORCE DEVELOPMENT AT NEW COMMUNITY CORPORATION

Created in the late 1960s in the predominantly African American Central Ward of Newark, New Jersey, New Community Corporation (NCC) has grown to employ approximately 1,400 employees, making it by far the largest CDC in the United States and indeed one of the largest single employers in its city. Case writer Ann Griffin, a master's student in city and regional planning at Cornell University, notes that NCC's operations now include housing development and management, day care and other human resources programs, nursing homes, a supermarket and numerous smaller retail businesses, and a growing complex of ET and workforce development (WD) activities.

Despite its focus on self-reliance, NCC's leadership learned at the earliest stages of operations how to network beyond the neighborhood to attract supporters who would become sources of social as well as financial capital. As early as 1968, NCC leaders spoke frequently at church groups and other gatherings—even in remote suburbs of Newark—recruiting supportive women as "Operation Housewives" volunteers. These volunteers joined colleagues from

NCC's Central Ward neighborhood in operating a thrift shop to raise funds for the CDC's first Babyland nursery, established in 1969.

Today, NCC's local networking efforts have lured leading corporate, education, and trade union figures from throughout the region to serve actively as members of the New Community Foundation, which operates separately from NCC's board of directors (and is quite distinct from the Industrial Advisory Board still being refined under an ongoing collaboration with San Jose's CET). Among the foundation's activists are high-ranking executives from such Newark area companies as Johnson and Johnson, Prudential, Bell Atlantic, Lehrer, McGovern and Bovis, Paine Webber, Deloitte and Touche, Bozell Worldwide, Supermarkets General (Pathmark), and the local electric power company as well as business leaders from banks, colleges, and universities. The New Jersey AFL-CIO is also represented on the foundation.

NCC's networking efforts regularly extend beyond the region to Washington, D.C., where its leadership is active on boards of advocacy groups focusing on community reinvestment, credit unions, and the more general agenda of CDCs as a national movement. Formal alliances extend west to California, through its partnership with the CET, and east across the Hudson River to the World Trade Center in Manhattan, where NCC works with the Port Authority of New York and New Jersey. Recently, NCC leaders traveled even further east, to central Europe, where leaders established relationships in Prague, Budapest, and other capitols in that region.

Who Are the "Customers"? Organizational Capacity-Directed, Neighborhood-Directed, and Regionally-Directed Training

More has probably been written about NCC than about any CDC since the Bedford-Stuyvesant Restoration Corporation. Here, we follow Ann Griffin in focusing on NCC's ET and WD activities, especially their growing engagement in external network collaborations.

Throughout its existence, NCC has emphasized the value of skill upgrading and lifelong learning for its employees. To implement these objectives requires continually finding additional sources of job openings. As the CDC expanded its project activities beyond housing, each potential venture was evaluated to determine whether it would create employment opportunities for residents of the Central Ward. As the number and diversity of NCC's ventures grew, the organization began to reach out beyond the immediate neighborhood for people who could impart valuable experience to other members of NCC project personnel. Individuals with police enforcement background were brought in to serve as leaders for NCC's security division. People with social service programmatic backgrounds were brought in to operate human services divisions. Those with engineering construction and building systems skills became physical plant and operating systems managers. Wherever feasible, NCC has endeavored to promote local residents within its own workforce. It quite deliberately never limited itself to employing only local residents, however.

Senior staff at NCC like to evaluate assets that the CDC owns, such as its fleet of delivery and passenger vans, and then devise a training program for young workers to retrofit and rehabilitate such vehicles. Furthermore, where such skills might allow for launching a viable entrepreneurial endeavor, the CDC staff considers that possibility—for example, creating a program to repair vans owned by other nonprofits and public agencies in the greater Newark area.

Over time, supervisors and trainees from these various programs sometimes move to private or public companies in the greater Newark area. These individuals now routinely return to NCC to recruit additional employees with comparable skills to fill new openings with their employers, who are dispersed throughout the area. In this manner, NCC "seeds" the region with its own former employees and, in the process, reinforces social networks that enhance the job access of the next generation.[10]

NCC has also attempted to seed the region with potential entrepreneurs who started at the CDC as trainees. This has occurred, for example, as trainees working on commercial food initiatives have occasionally moved from working at NCC restaurant businesses to becoming independent caterers on their own. In a few instances, NCC has provided tuition and related costs for its own personnel to attend training programs for future caterers. Similarly, NCC has underwritten the training of franchise managers and key employees with cooperating companies such as Mailboxes, Etc. and Dunkin' Doughnuts. For example, senior personnel of these NCC-owned ventures were sent for training to Massachusetts and California, respectively, in a manner consistent with their franchising agreements with these two prominent companies.

The CDC anticipates that its own managerial personnel from these franchises will eventually become owners of their own units. NCC's philosophy is to perceive such personal success stories not as a problematic loss of skilled managers but instead as an opportunity to promote a cadre of new managers from among the well-trained personnel of the NCC-owned franchise unit.

NCC's Employment Services Center

In 1984, two devoted NCC members, John and Mary Binns, established the Employment Services Center at NCC. Case writer Griffin points out that the founders of the center were particularly inspired by their commitment to social justice and a religious faith of personal sacrifice for a community cause. With limited resources and staff, the Binns nevertheless managed to expand NCC's job brokering and placement activities to an average of 500 clients finding employment per year by the 1990s (the rate was almost twice that when the Newark economy was, for a time, expanding during the 1980s).

As NCC's Employment Services Center staff has grown to a contingent of 10, the number of employers with whom the center's job developers communicate has come to exceed 1,000. The staff report that their most effective job developers are those individuals whom private-sector firms have come to trust when referring candidates for employment. The Employment Services Center screens appropriate candidates for available positions and provides or directs them to job readiness training where needed. Employers (who are charged no fees) regularly cooperate with job developers by providing written evaluations on those whom they have hired, by responding to telephone postplacement follow-ups, and by inform-ing NCC staff when they believe that personal or family interventions might help a referred employee deal with problems interfering with job attendance or punctuality.

Employment center initiatives now receive funding from a variety of sources, including the United Way, the State of New Jersey, and federal programs including JTPA. In addition to placement activities, the Employment Services Center also operates a state-sponsored welfare-to-work program and youth training programs. NCC employ-ment counselors assist clients with a range of support services, including child care, medical care, and even affordable clothing and furniture. When needed, referrals to legal services, counseling, and health services are afforded to clients. Like the best of the training programs throughout the country, however, NCC counselors do not stop at merely informing clients about services. They regularly make phone calls directly to service providers, on behalf of clients, and then follow-up to confirm that the service providers actually re-sponded to clients' needs. In other words, they act as intermediaries, brokers, and advocates.

The NCC Collaboration With the CET

In 1992, NCC decided to expand its placement and work-orientation activities to the provision of a range of specific occupa-tional skills known to be in demand in the region's private sector. As we have seen, the CDC had for some time provided training for its own internal workforce, in areas such as construction, facilities maintenance, child care, and security, and was also providing or brokering limited training for outside employers, mainly in human services. NCC leadership determined that the needs of the commu-nity required a training model that could more systematically assess job opportunities in the private sector and expand the organization's capacity to offer training, all while ensuring that local residents could be provided the necessary skills to meet the requirements of private firms in the greater metropolitan area.

NCC Executive Team member Florence Williams had become familiar with the CET while working as New Jersey's Assistant Commissioner of Human Services. NCC's leaders were particularly impressed with how CET used its IAB to network with senior corporate officials from the private sector. The IAB vehicle was also seen as a valuable tool for informing training program designers and instructors about changing technologies and skill requirements. Building on the preexisting New Community Foundation, Williams and NCC founder and Chief Executive Officer, Monsignor William Linder, hoped the NCC would get better at ensuring that skills taught through NCC's programs would continue to be in demand by private firms.

In 1992, Newark was designated as a site for the U.S. Department of Labor's national CET replication—one of the handful (along with New Haven and a few others) that have been set up as true peer-to-peer inter-CBO networks, with the city's official JTPA agency in the background.[11] The collaboration had a rocky beginning, and some of its initial staff members are now gone. Despite its admiration for CET's operating approach, the leadership of NCC had to negotiate sensitively with CET's own leadership to vary certain programmatic approaches and operating styles. For instance, for several programs New Jersey state legal constraints preclude the open entry-open exit operating feature, which is a much admired aspect of CET's learning approach. Other management prerogatives also came into consideration from time to time when operating priorities, such as interaction with local community colleges, were less consistent with the conventional CET approach (CET tends to view community colleges as rival training vendors and only occasionally works with them on the West Coast).

By contrast, for a number of years, NCC has had impressive relationships with leading academics from a number of universities and community colleges. In addition to lifelong learning skills, NCC trainees have also received intensive training in environmental sciences, math, and chemistry from a rotating roster of visiting professionals provided by Bloomfield Community College through funding from the U.S. Department of Defense. The Center for Environmental Engineering and Science at the nearby New Jersey Institute of Technology (NJIT) has a long-standing relationship with

NCC training personnel and regularly fosters collaboration and information sharing between NJIT instructors and NCC-CET trainers.

NCC's partnership with CET has also afforded the latter access to NCC's day care programs at its award-winning subsidiary, Babyland Nursery, Inc. For example, a partnership was recently established with Essex County College, which provides instructors for a child development associate training course with on-site training and subsequent employment at Babyland facilities. Some instructors from Essex had had relationships with NCC's Babyland programs for a number of years prior to the arrival of the CET program. Indeed, Essex County College President Zachary Yamba, the Chancellor of Seton Hall University, Reverend Thomas R. Peterson, and President Francis J. Mertz of Farleigh Dickinson University are all members of the Board of the New Community Foundation, an organization of corporate leaders, educators, and accomplished individuals who develop resources for NCC and generally act as "friends of the family."

The point is that, through this peer-to-peer collaboration between what are the largest CDC and the most effective "alternative" ET vendor in America, organizational learning is taking place in both directions. Another example of this is the June 1996 retreat for the staffs of all the East Coast CET replication sites. The meetings were held at NCC.

There is an interesting nexus in how NCC integrates economic development, workforce development, and seeding the regional economy. No other CBO in the country does it quite like NCC does, and it deserves closer attention. For example, as NCC embarks on its commercial vehicle repair and retrofit initiative, a foundation board member and owner of New Jersey's largest auto sales firm has pledged $100,000 worth of equipment and future job openings (he also contacted Edsel Ford and obtained a $100,000 donation from the Ford Motor Company). The new commercial vehicle facility was officially opened and dedicated in December 1996.

Although the CDC is enthusiastic about its ability to connect to such resource providers, however, it perceives the new initiative as much more than a job training and placement program in vehicle repair and retrofitting. NCC is particularly anxious to expose youths in this training initiative to the use of computers (particularly CAD/CAM and diagnostic equipment) in a nonclassroom setting with instructors who also serve as fellow workers within the new venture. Most of these young trainees will ultimately graduate to firms—including those in the health and transportation services sectors, for example, at the region's seaports and airports—that use computer applications for diverse technologies but not necessarily related to vehicle repair per se.

Thus, the initial vehicle retrofitting enterprise-specific training program becomes a means to a larger end or, to use economists' jargon, produces a "joint product": a potentially profitable export-oriented business venture and a flow of skilled workers who can also be "exported" beyond the inner city.[12]

Luring Instructors From the Private Sector

One important parallel between the training approaches of NCC and CET is the recruitment of instructors from the private sector. The NCC-CET Community Health Worker Training (CHWT) program clearly demonstrates the value of instructors being able to reach out to colleagues in private hospitals and other firms, sometimes by virtue of having been recruited from them. Case writer Griffin reports that CHWT trains individuals in managed care, prenatal and postnatal care, domestic violence, HIV/AIDS, and sickle-cell anemia. Trainees acquire health counseling skills and learn referral techniques involving public health and medical facilities as well as social service programs. With initial funding from the Robert Wood Johnson Foundation and additional support from the Pfizer Foundation and Project Hope, each trainee receives extensive individual job development and placement counseling. Counselors meet with trainees on a weekly and biweekly basis throughout the training course and submit monthly monitoring and progress reports.

To date, all 35 graduates of the program have succeeded in landing positions, with starting salaries averaging $21,000. The successful placement of all graduates was not accomplished merely by providing first-rate training. Graduation speakers from the health industry (as well as the instructors themselves) described to colleagues the enthusiasm and skill generated by the program participants. As in the CET and QUEST cases, so it is with NCC: These ongoing associations with business leaders and instructors become the networking resources that graduates find so valuable in landing permanent jobs.

Merging NCC's Proliferating Training Activities

As they evolved, NCC's Employment Services Center, the CDC's internal workforce programs, and the NCC-CET undertaking operated with separate management, funding, and even facility locations. Sometimes, staff in NCC's Human Services Department, directed by Florence Williams, received little information regarding programmatic developments in the CET program. On other occasions, NCC and CET staff were unaware of vital corporate contacts already developed and maintained by organizers and job developers at the Employment Services Center.

As a result, in 1995 a 10-person design team was formed by Monsignor Linder that was composed of senior NCC staff and charged with analyzing alternative approaches to streamlining and consolidating the CDC's proliferating employment and training systems. By this time, ET and WD were being viewed as a cornerstone of the NCC mission. By June, NCC had put in place a new, overarching Workforce Development Department (Figure 5.13). Serving as a facilitator during this redesign process was Bill Tracy, a former director of the New Jersey State Commission on Employment and Training. NCC's Kathy Spivey, a human resources expert whom NCC recruited from the McDonald's Corporation, assumed the directorship of the new department and reiterated that workforce development would be regarded as a central activity for the CDC rather than a sidelight to traditional housing and economic development undertakings. In the fall of 1996, Ms. Spivey was promoted to become Monsignor Linder's chief of staff, further signifying the enhanced role the CDC is assigning to its workforce development activities. Pat Cooper recently came on board to manage the Department of Workforce Development.

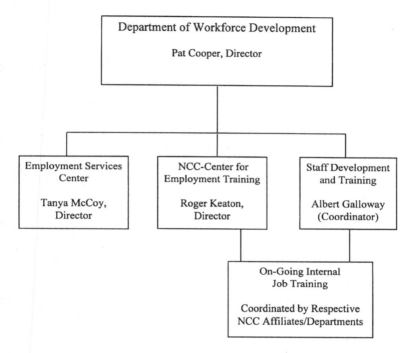

Figure 5.13. New Community Corporation Job Training and Job Placement Programs in 1996

NCC-CET is now the primary training delivery component within NCC for all forms of internal and governmentally funded training programs. Most important, barriers between operations are being overcome. Coordinating and planning meetings have become commonplace. Teams are being created to focus on youth, community health, the general population, and staff development. This coordination effort remains a work in progress.

Clearly, the state of New Jersey recognizes the value of NCC's ability to link its programs and trainees to private-sector firms. Monsignor Linder, Spivey, and Williams continue to serve on a number of state advisory boards and task forces, under both Democratic and Republican governors. CET also demonstrates this extraordinary staying power, without ever being "apolitical." Indeed, CET's former director of (California) statewide projects, Carlos Lopez, now serves as a deputy commissioner at the state and is in charge of ET programs. In Texas, QUEST is almost as favored by the current governor, Republican George Bush Jr., as it was by Democrat Ann Richards.

CDC-Based ET Networking and Community Building

All this acknowledged success at creating, entering, and sustaining networks with influential private-sector firms, governments, and other organizations (such as CET) does not, by itself, answer the question of the extent to which NCC has had a measurable impact on the lives of the residents of the Central Ward of Newark. Indeed, one hears—even from the staunchest allies—the constant concern that the CDC movement as a whole has turned into landlords and social service agencies run by professionals from outside the neighborhood and that, from their race and gender composition, many CDC boards do not represent their largely minority and female local constituencies.[13] One new study raises concerns that CDCs such as NCC do a better job with "brick and mortar" (housing and commercial) development than with "community building" per se. In particular, conclude these researchers, NCC has been far more successful at representing and mobilizing community interests than at creating participative structures per se (Briggs & Mueller, 1997). This is an old tension in the community development movement.

We put these concerns to the leadership of NCC. They responded by pointing out that the majority of their employees are drawn from the Central Ward (and are African American), and that NCC services—including housing, child care, the retailing of quality but inexpensive food, and employment training—are offered mainly (although deliberately not exclusively) to local residents.

That "outsiders" participate in the affairs of NCC—as staff, customers, service providers, and advisers—is seen by leadership as a strength—a key to understanding the CDC's remarkable survival and growth during a quarter of a century. These external relationships are said to bring into the community resources that would not otherwise have been available, not only to the organization but also to residents.

Three decades ago, when theories of community economic development were at the forefront of American social policy concerns, it was well understood that the greatest challenge for indigenous organizations would be to achieve scale, effectiveness, internal organizational capability, and political influence without sacrificing the grounded commitment to radical social change along the way. Probably no CDC or other CBO in America has achieved those objectives, while "keeping the faith," more than NCC. This is the larger context within which to understand, appreciate, and (eventually) formally evaluate NCC's workforce development networks.

CHICAGO'S BETHEL NEW LIFE, INC.

Gloria Cross, PhD candidate in public policy at the University of Massachusetts-Boston, is the author of the fourth case study. She writes:

> Founded in 1979 by Bethel Lutheran Church, Bethel New Life, Inc., is a community development corporation based in Chicago. Bethel began as a CDC, with an emphasis on housing, serving the predominantly African American neighborhood of West Garfield Park. With more than 400 employees and an annual operations budget of over $6 million, Bethel is now a multifaceted CDC with programs in housing, health and human services, senior services, community building ("neighboring") initiatives, jobs, and economic development.
>
> From its inception, Bethel has utilized informal relationships and connections with individuals and organizations to facilitate access to resources. Whether through volunteer tutors from suburban churches or weekly $5 donations from church members to renovate its first "three-flat," human and financial resources have been attained [in this way from] both within and outside the community.

Bethel's first formal, contractual employment training network, the Certified Nursing Assistant program, was launched in 1987 (one of its graduates was prominently featured in the film *Hoop Dreams*). Since then, the CDC has entered into or sought out a growing number of interorganizational collaborations, involving training in health and office skills, welfare-to-work programs, environmental and industrial policy, and youth entrepreneurship initiatives. Some of these networks are local, some are regional, and at least one (the Youth Enterprise Network) is national in scope.

Although the president, Mary Nelson, is supremely well connected herself, Bethel is widely considered within Chicago as tending to be more collaborative with outside and mainstream organizations than with CBOs located inside or near Garfield Park. This is both a strength and a weakness, argues Cross (and in any case is common to other large CDCs, such as NCC). The external ties clearly bring resources that would not be available otherwise into the neighborhood. Many opportunities to partner locally, however, are overlooked (or not acted on), such as the chance to work with Suburban Job Links, a nationally known West Side-based reverse-commuting program that works with employers, city councils, and transportation agencies to get inner-city residents out to suburban job sites, and with the national demonstration program of which Job Links is an integral part: the Bridges to Work project of Public/Private Ventures, Inc.

To be sure, Bethel is amply stocked with well-connected executives. For example, Lawrence Grisham, Senior Vice President for Operations, also serves as vice president of the Chicago Rehab Network. Mildred Wiley, Director of Community and Education Programs, is also president of the Neighborhood Capital Budget Group, an alliance of CBOs that advocates for and monitors the city government's physical investments in infrastructure in the neighborhoods. Sheila Radford-Hill, President of Bethel New Life's board of directors, is also (among many other things) the former director of the Chicago chapter of the national network, Cities in Schools, with its many local affiliates. These are individual activists, however, and these remain largely personal connections and not formal network alliances contractually linking Bethel, as such, to these other inside-Chicago organizations.

The Central Role of Religious Institutions

Bethel's origins as a black CDC are traceable to the white flight of the early 1960s. Between 1960 and 1965, West Garfield Park changed from being 85% white to 85% African American. According to the 1990 census, 45% of the area's residents received Aid to Families with Dependent Children (AFDC), with half of those recipients on the rolls for 4 years or more. Jobs, especially in manufacturing, continue to disappear. A significant fraction of the population still cannot read at a sixth-grade level. West Garfield Park is clearly an area in great distress.

Just as CET, QUEST, and NCC were all created out of, by, and through the actions of church-related organizers, so was Bethel New Life. Indeed, in this case, the behind-the-scenes role of the church continues to be a force today, nearly two decades after the founding of the CDC.

Bethel Lutheran Church was first established in 1890 by German immigrants. During the 1960s, as in so many American cities, the color of the parishioners changed with that of the population. (White) Pastor David T. Nelson was installed to lead the church into the new era. An interdenominational group of four ministers was formed to work cooperatively, drawing tutors into West Garfield Park from suburban churches, and Bethel created its own Christian school (preschool through eighth grade). At one point, the interdenominational group included as many as 13 churches and eight denominations, with Mary Nelson (Pastor Nelson's sister) as director of planning and development. This first interorganizational network lasted 15 years. Eventually, issues regarding corruption in the utilization of funds were raised and could not be resolved, and the network collapsed as member churches withdrew. Out of this failure, Bethel Lutheran Church then tried a different direction, creating the CDC, Bethel New Life, in 1979.

Bethel New Life, Inc., continues to maintain strong informal and formal ties to the church. For example, Mary Nelson, the CDC president, continues as church organist. Eight of the 15 CDC board members are parishioners. Bethel Lutheran Church has mortgaged itself five times to support CDC projects, which also continue to receive funds from the Evangelical Lutheran Church of America. Many staff members we interviewed are active in the church. No other group we have studied is so deeply embedded in an ongoing way within the context of an organized religious institution as is Bethel New Life.

Workforce Development at Bethel

In 1984, the CDC created its Bethel Employment and Training Services department (BETS). Initially financed by grants from Chicago United Way, it offers literacy, job readiness, and skills training. According to the current director, Cedric Melton, until recently, Bethel relied entirely on grants and contracts from foundations, the church, and the Illinois Department of Public Aid, avoiding any reliance on the city's JTPA agency.[14]

On both a walk-in and referral basis, BETS provides residents with educational and skill assessment and testing and matches them to available training or placement opportunities. The CDC is also engaged in pressing employers to create more such opportunities for its client population. Bethel conducts considerable postplacement counseling. Using a new computerized database—among the best we have seen anywhere—BETS is quite able to meet federal government job training requirements for tracking all trainees it places for up to 90 days. All services are provided to area employers at no cost. On the basis of nearly 10 years worth of evidence on performance, consultants to the Enterprise Foundation recently ranked BETS as one of the 3 most successful ET centers (in terms of placement rates) of 14 such agencies throughout the country that have been supported by that foundation (Gittell & Wilder, 1995, p. 16).

Within Bethel New Life, as with NCC (but on a far smaller scale), BETS also works closely with the in-house human resources office to provide screening and training for employees of the CDC's own housing, industrial, and social service ventures.

Locating job openings outside of Bethel has become increasingly difficult during the past few years, with continuing deindustrialization, heightened interracial and interethnic competition for what jobs do exist, and as more employers get "lean and mean," being more inclined to offer mainly part-time and contingent work (Harrison, 1994; Tilly, 1995). To get area employers more involved in the concerns of Bethel and its clients, the CDC recently organized an Employment and Training Advisory Panel composed of local companies and social service agencies. A few of the companies represented on the panel have directly hired people trained by (or through) Bethel New Life.

Nevertheless, although superficially similar to CET-OIC (and QUEST) industrial advisory boards, we found no evidence that the member firms and agencies were consistently hiring trainees from West Garfield Park, let alone providing instructors or equipment to BETS or to the vendors with whom BETS contracts. Perhaps the CDC could make a concerted effort to pursue such an objective, especially because there is at this time no effective CET presence in Chicago.[15]

Employment Training and Workforce Development Networks

Bethel has started or joined a great many such networks over the years, especially since 1990 (see Table 5.4). Here, we discuss only four of these (for more details, see the actual case study by Gloria Cross).

The Youth Enterprise Network (YEN), launched late in 1990, has proven to be the most durable of these network alliances (Fitzgerald, 1995; Harrison, Weiss, & Gant, 1995, pp. 33-34). YEN is the Chicago site of an eight-city national demonstration project initiated from Cambridge, Massachusetts, by the Center for Law and Education. It seeks to more closely connect (and, in the process, reform) vocational education in three West Side high schools with community economic development principles, using funding from the U.S. Departments of Labor and Education, under the federal Perkins Act. Students in YEN learn to collect and analyze data on the local economy, develop business plans, and actually operate small in-

Table 5.4 Bethel New Life's Employment Training and Economic
Development-Related Networks

Program	Year Established
Recycling Center	1984
Certified Nursing Assistant Program	October 1990
Bethel Self-Sufficiency Program	August 1991
Material Recovery Facility	1992
Environmental Jobs Initiative	July 1994
Operation Jump-Start	August 1994
Bethel Training Institute	August 1994
Office Skills Internship	October 1994
Urban Engineers Program	October 1994
Industrial Outreach/Technical Assistance	1995
Industrial Development	1995
National Foundation for Teaching Entrepreneurship	June 1995

school businesses that serve not only the student population but also areawide residents.

Under Mildred Wiley's leadership, YEN (originally directed by Sheila Radford-Hill, who is currently the president of Bethel New Life's board of directors), was designed from the beginning to engage parents and other residents in program design and management of what has become a peer-to-peer network of its own, which includes Bethel, the three high schools, the Chicago office of Cities in Schools, a community college, two universities, and another citywide community development network, the Chicago Workshop for Economic Development. Moreover, YEN has clearly had as a longer range goal the promotion of community building, mobilizing parents around other aspects of public school policy, and providing training for area residents interested in joining local school councils. In other words, it is aimed at changing the public school system. In 1996, the Chicago Empowerment Zone contracted Bethel to create a Youth Enterprise Institute for all the high schools in the zone.

A number of projects have also evolved from Bethel's ongoing relationship with nearby Argonne National Laboratory. Argonne, which is managed by the University of Chicago, is the U.S. Department of Energy laboratory that, with its staff of more than 4,000 persons and an annual budget exceeding half a billion dollars, is one of the premier sources of technology transfer and research and development in America. A 1993 bus tour of the West Side brought Argonne's director, Allen Schriesheim, associate director Harvey Drucker and Schriesheim's assistant Norm Peterson into the community, and they and Mary Nelson hit it off (it turned out that Drucker and Peterson had grown up near Garfield Park). In the space of 2 years, an Argonne-Bethel partnership (which also includes such existing Argonne collaborators as Ford, GM, IBM, and Caterpillar) has created training opportunities in environmental (lead paint removal and hazardous waste treatment) assessment and remediation, industrial waste recycling, office skills, and the dramatic Urban Engineers Program (whose manager, inside Bethel, is also Mildred Wiley).

Urban Engineers is actually an alliance among Bethel, Argonne, the University of Illinois, the Illinois Institute of Technology, and the Urban League. It brings environmental engineers—often dressed in their high-tech "clean suits," looking like heroes out of a Hollywood film—into West Side public schools to give career "visioning" workshops for students in Grades 4 through 12 (and, sometimes, their parents). The first summer internships at Argonne, Hewlett-Packard, and other participating companies were provided in 1995 (Argonne National Laboratory, 1995). Facing a severe budget crisis, which threatened the continued funding of Urban Engineers, Bethel and Argonne created a planning process to develop a 5-year plan that emphasizes jobs, sustainable community development, energy efficiency, and environmental policy. These have been Argonne's objectives since the downsizing of the national military and atomic energy budgets; Bethel has been involved with environmental issues for some time.

Although Garfield Park, like the rest of Chicago, has undergone unrelenting deindustrialization during the past 20 years, it remains home to approximately 40 small and medium-sized companies, mostly in the machine tool industry, as well as to a few large manufacturers. Bethel recently joined forces with Argonne and the Chicago Manufacturing Center (CMC), the latter being one of 44

such centers within the national Manufacturing Extension Partnership of the federal government's National Institute of Standards and
Technology. Teams visit local manufacturers to "benchmark" their
operations and recommend specific industrial extension services
that might enhance their competitiveness, thereby retaining jobs on
the West Side. Bethel's relationship with Argonne is quite formal;
with the others (CMC, the Austin Labor Force Initiative, and Shorebank Enterprises) it is more informal.

A fourth network shows signs of evolving into something more
substantial. As we have seen, Bethel New Life has for 10 years been
engaged in training home health care workers. In 1996, the CDC
received a $500,000 grant from the Job Opportunities for Low
Income Individuals Program of the Office of Community Services
(OCS) within the U.S. Department of Health and Human Services—a
small agency that has, since the 1970s, been a reliable and consistent source of modest funding for CDCs across the country. OCS
wants Bethel to focus its home health care program on training and
finding or creating job ladders for 80 welfare recipients. By including
AIDS patients and veterans within its clientele population, Bethel
has also been able to gain access to managed care contracts,
thereby further expanding its revenue base.

In operating the program, Bethel collaborates with the Illinois
Department of Public Aid, local hospitals, and the city college
campuses that offer training programs for certified nurses' aides and
related occupations. The CDC also provides or arranges for day care,
transportation, housing assistance, and other family services. Even
entrepreneurial spin-offs in the health care field are contemplated.

The Paradox of Organizational Capacity

YEN (the national network into which Bethel was invited by
Cambridge-based educator Larry Rosenstock) has flourished. The
home health care training network looks promising. Urban Engineers is in financial difficulty. Other networks created by Bethel,
such as its materials recycling facility, linked to Helene Curtis and
several recycling specialty firms, and a short-lived collaboration with
IBM to provide computer-related training through a new Bethel

Training Institute, have been closed down. In contrast to NCC, whose collaborative initiatives seem generally to expand over time through careful nurturing, problem-solving, and occasional crisis intervention from the highest levels of the CDC, Bethel New Life displays a history of launching many network alliances but being unable to sustain more than a few of them over the long run.

In theory, network engagement should have increased Bethel's capacity to meet local needs by gaining it access to ideas, resources, and contacts. The case reveals, however, that even for such large organizations as Bethel, strategic alliances and other collaborations are not free or even cheap. Bethel has tended to staff the management of its networks by turning over initiatives begun by the president and a few other senior executives—often serendipitously—to existing personnel, adding to their already overflowing portfolios, rather than by expanding its internal capacity and otherwise altering internal practices to better fit the changing demands made by network partners.

Thus, Bethel illustrates that networks can add additional layers of complexity to already overburdened CBOs, making it more difficult for existing staff to do their jobs. Cross also thinks that the devotion of staff effort to managing external networks can (and, in the case of Bethel, does) make it even more difficult for residents of the community to become involved in decision making about workforce development and other activities.

These are correctable problems, especially for a CBO with the staff depth, skills, and experience of Bethel New Life. What is probably needed most is a stronger, even philosophic, commitment by the president and other members of the CDC's leadership to staff development, greater delegation of responsibility, and the will to sustain network alliances once they have been set in motion. "Sustaining alliances" does not always mean "running programs." It means that whichever members of the network actually manage the operations (in this case, WD), the CDC does not simply initiate and then walk away but rather develops the capacity to nurture the initiative and continues to hold it accountable to neighborhood concerns.

A RURAL-SMALL CITY CDC NETWORK:
MAINE'S COASTAL ENTERPRISES

The one case that began as a rural and small-city hub-and-spoke model is Coastal Enterprises, Inc. (CEI). Still based in the small coastal community of Wiscasset, Maine, but active along the coast of the country's most northeastern state, and now working in cities such as Portland, CEI differs from traditional, urban CDCs by deemphasizing real estate development and focusing instead on connecting its venture capital investment initiatives to job placement, job training, and the provision of support services to individuals and cooperating businesses.

Developing Business-Like Contractual Relationships

CEI views its strength as being a lender of last resort to emerging enterprises that are willing to enter into formal, legally binding agreements to hire low-income individuals referred by the CDC. Coastal Enterprises conducts research to determine which types of businesses are most likely to offer career ladders, as well as entry level positions, to individuals needing assistance to enter the private-sector workplace. CEI's networking efforts are an intrinsic part of its ability to refer qualified workers who are well trained prior to employment or on the job. Its ability to achieve success with respect to job-retention levels is also enhanced by CEI's close working relationships with state, college, and private job training agencies and postplacement support programs. Although venture development linking with job contracts is CEI's most significant activity, the CDC also provides financial support and technical assistance to collaborating social service organizations. It also has a range of affordable housing resources that facilitate home ownership among its targeted low-income populations.

CEI was incorporated as a CDC in 1977. Although it was established as a rural CDC servicing southern mid-coastal Maine, its geographical area has grown to include Portland as well as other urban areas. Its venture capital fund covers the entire market of Maine. CEI has 40 full-time staff.

In the past 18 years, it has slowly introduced new and innovative programs to meet the needs of its constituents. It has evolved from supporting natural resource-based industries, such as fisheries and small-scale woodworking enterprises, into a complex organization that provides a wide array of financing programs, technical assistance, and social services.

CEI maintains a one-stop business development center where businesses and aspiring entrepreneurs obtain access to loans, venture capital funds, SBA subordinated long-term loans, and small enterprise loans. The development funds are complemented with a full range of technical assistance. The businesses that seek financial assistance from CEI enter into a binding contract to hire low-income people. The businesses have to create one job for every $15,000 of CEI's funds. As of 1994, CEI provided $24.5 million in financing various business enterprises. These funds have leveraged an additional $70 million from private financial intermediaries. CEI's accomplishments include the creation of 5,000 jobs, providing financing to 400 enterprises, assisting in the establishment of 60 day care centers, serving 10,000 individuals or businesses, and financing 175 housing units.

With limited financial investment and personnel, CEI has managed to bring together numerous employment-related organizations. In this regard, it is considered a team builder because it has successfully transcended territorial conflicts that previously inhibited antipoverty organizations in Maine from working for a common goal: community colleges, women's development centers, state and federally funded employment programs, and other support organizations.

The various organizations bring to the table specialized services and assistance to make the models work. They tailor training programs to fit employers, screen potential candidates for the training programs, provide day care and transportation assistance for the candidates during the training programs, work with the candidates while they are on-the-job training, establish a smooth transition period for candidates and their employers, and provide ongoing postemployment follow-up and mentoring.

Overcoming Financial Constraints

The leadership made an early decision that reliance solely on governmental grants would constrain it from achieving the level of scale and investment penetration to cover critical overhead. Although its initial emphasis was mid-coastal Maine, CEI decided that its development funds needed to operate with a minimum capitalization level of $5 to $7 million dollars to generate sufficient interest-earning revenue to cover the level of technical assistance and postloan support. CEI recognized that technical assistance at the pre- and postloan stages would prove critical to nurturing business and individuals during periods of varying economic cycles. As a result, CEI assumed operational responsibilities for maintaining Small Business Development Centers for the U.S. Small Business Administration and the state of Maine. These responsibilities led the CDC to open field offices in Augusta and satellite appointment locations in five other Maine counties. Numerous program activities also take place frequently in Portland.

Our case writer, Dr. Mulugetta Birru, Executive Director of the Urban Redevelopment Authority of Pittsburgh, analyzes CEI's networking activities in the context of its desire to be a change agent for low-income communities in Maine. Birru points out that CEI's regular interactions with job training programs, human services agencies, banks, employer groups, and political actors are a critical factor in ensuring the successful placement of low-income people in jobs created by firms in which CEI invests. CEI has come to perceive its networking resources and responsibilities as requiring its leadership to regularly pursue policy and programmatic changes at the state and federal levels.

*Targeting Opportunities to Facilitate
Job Creation for Poor People*

Given its emphasis on serving low-income individuals with job creation, CEI found it desirable to create a Targeted Opportunities

Department. This division of the CDC holds the other operational departments accountable to the job creation mission of the CDC. Prior to loan and investment commitments from the various CEI portfolios, the Targeted Opportunities Department gauges the job creation opportunities and other social benefits that might flow from the business being funded. With each venture finance application, this department prepares profiles of potentially available positions in the company. Such profiles evaluate job skill requirements, experience needed, salaries and wages initially available, and career ladder opportunities. Using this information, employment specialists in the Targeted Opportunities Department communicate with the extensive network of service providers funded through federal JTPA and federal welfare support programs. The director of CEI's Targeted Opportunities Department regularly serves on the boards or pertinent committees of these service providers. In addition, the department coordinates postplacement services to new employees to remedy any conflicts that may have arisen during the first year of employment. These postplacement service providers come from both within CEI's staff and, more often, from the diverse service providers with whom CEI networks.

Among the key state and local programs CEI regularly works with is the state's initiative, Additional Support for People in Retraining and Education (ASPIRE). ASPIRE provides welfare recipients with life-management skills, GED preparation, ability assessment and testing, specific skill training, continuing education, job search and placement, on-the-job training, and supportive services such as child care, transportation resources, and even required tools and uniforms. Maine also has a number of county-level workforce development centers that are funded by JTPA and operated primarily by nonprofit organizations. Workforce development centers often serve broader categories of the population than ASPIRE, which targets its resources exclusively to AFDC clients. Among other critical nonprofit programs regularly interacting with CEI staff are the Maine Centers for Women, Work, and Community, which target women who have become heads of households and are compelled by law to enter the workforce. Two related entrepreneurial programs with which CEI regularly interacts include the Maine Displaced

Homemakers program and the Women Business Development Corporation. CEI also collaborates with adult education training initiatives at Lewiston College and the University of Southern Maine.

CEI-Initiated Employment Programs

Coastal Enterprises has also taken advantage of federal resources to create two targeted employment initiatives. Project JUMP (Jobs for Unemployed Maine Parents) targets its support efforts toward job access for unemployed, two-parent heads of households. This program focuses on York County, an area with some of Maine's greatest unemployment problems and AFDC caseloads. The county has also been afflicted by the closure of important military installations.

Project SOAR (Structure Opportunities for AFDC Recipients) is another program through which CEI attempts to target opportunities to mothers receiving welfare in Androscoggin County and, in particular, Lewiston, Maine's second largest city. SOAR receives federal support from the U.S. Department of Health and Human Services Office of Community Services and matching funds from the state of Maine and local banks. This initiative represents a collaborative effort between CEI, the Maine Department of Human Services, the Displaced Homemakers Program, and the Women's Business Development Corporation. Using support services from the collaborative entities, CEI provides loans, technical assistance, and business counseling to AFDC recipient mothers who are capable of pursuing self-employment.

Convincing the Private Sector That "Unemployables" Can Do the Job

Wages in CEI-assisted businesses average approximately $9 an hour. Numerous assisted businesses report that they regularly provide health insurance, child care assistance, educational tuition reimbursements, and profit-sharing with their employees. CEI's research division also reports that employers and supervisors confronted their own initial skepticism about employing low-income workers. The same employers now acknowledge that low-income workers employed because of CEI loan commitments have integrated into their workforce with few significant problems. The postplacement interventions by CEI personnel and collaborating organizations continue to be acknowledged as an important resource for ensuring the productive transition of many workers from welfare to sustained careers.

NOTES

1. The case of CET is authored by Dr. Edwin Melendez, Professor of Economics and Director, Mauricio Gaston Institute for Latino Community Development and Public Policy, University of Massachusetts-Boston. The study of Project QUEST was developed by Dr. Rebecca Morales, Associate Professor of Urban Planning and Policy and former executive director of the University of Illinois-Chicago's Center for Urban Economic Development. The case on NCC is written by Ann Griffin, a master's student in the Department of City and Regional Planning at Cornell University. Gloria Cross, doctoral candidate in public policy at the University of Massachusetts-Boston, is author of the case on Bethel New Life, Inc. Coastal Enterprises was studied by Dr. Mulugetta Birru, Executive Director of the Urban Redevelopment Authority of Pittsburgh. The original cases can be purchased directly from the Economic Development Assistance Consortium, One Faneuil Hall Marketplace, Boston, MA 02109. Phone: (617) 742-4406; e-mail, MarkEDAC@aol.com.

2. Marcus Weiss is the author of the case on Chicago Jobs Council. The study of Pittsburgh Partnership for Neighborhood Development was written by John Metzger, Assistant Professor of Urban and Regional Planning, Michigan State University. David Sweeney, Executive Director of the Greenpoint Manufacturing and Design Center in Brooklyn, authored the case on the Business Outreach Centers.

3. Leslie Winter, former director of the Real Estate Institute of New York University and the Office of Real Estate in the City of New York's Office of Business Development, wrote the case on the Regional Alliance. The case on Lawson State Community College is authored by Dr. Joan Fitzgerald, Assistant Professor of Urban Planning at the University of Illinois-Chicago and Associate of the University's Great Cities Institute.

4. Technical evaluations of CET include Cave, Bos, Doolittle, and Toussaint (1993), County of Santa Clara (1974), Development Associates, Inc. (1971), Hershey and Rosenberg (1994), Hollister (1990), Kerachsky (1994), U.S. Department of Labor (1995), and Zabrowski and Gordon (1993).

5. Although the "official" JTPA-financed replication process continues, other CBOs throughout the country—for example, in New York City and in Oakland, California—are beginning to collaborate directly with CET, with the marriage-maker being the Local Initiatives Support Corporation.

6. Of course, as Melendez notes, a residential program such as Job Corps has other objectives than wage improvement and is also inevitably far more expensive.

7. Actually, the concept of a CBO developing close, continuing, stakeholding relationships to companies that will provide its trainees with jobs, or its minority enterprises with procurement contracts, has been a prominent characteristic of CET dating back to its OIC-SCC formative years in the late 1960s. This is not surprising given CET's origins as the West Coast branch of charismatic (and politically savvy) Philadelphia Reverend Leon Sullivan's Opportunities Industrialization Center movement. Indeed, workforce development professionals commonly speak of the "OIC model" of employer stakeholding. Melendez's case study discusses this history in detail.

8. The basic theses of the structuration school are set out in Giddens (1984).

9. This is also the period when CET became a kind of community development corporation, with property management, food stamp distribution, Montessori child care, migrant housing and hotel construction, janitorial, and auto repair and body shop enterprises. Even today, CET is engaged in building affordable housing, restaurants, and a mercado in downtown San Jose.

10. Our evidence on this is, at the moment, purely anecdotal. It is a question on which further research is necessary—and feasible. Such seeding would constitute an important indicator of a broader social impact of NCC on its region.

11. Although the "official" JTPA-financed replication process continues, other CBOs and local governments—for example, in New York City's South Bronx, Indianapolis, Kansas City, and Oakland—are beginning to collaborate directly with CET through the auspices of the Local Initiatives Support Corporation.

12. This idea of strategically exploiting possibilities for the generation of joint products (or, as they are sometimes called, "economies of scope" or "synergies") is important in the study of economic growth and development, generally (Romer, 1986). Once, it was much discussed in the field of community economic development but then became unaccountably neglected (Vietorisz & Harrison, 1970).

13. These characterizations are drawn from such basically sympathetic critiques as those offered by Berndt (1977) and Gittell, Gross, and Newman (1994).

14. "Hassles" with the Mayor's Office of Employment and Training (MET) are well-known in Chicago. Cross found many CBOs that do contract with the city as service providers to have a "love-hate" relationship with MET. This echoes a more widespread, national criticism by CDCs and other CBOs of the inflexibility of JTPA (see Harrison et al., 1995, pp. 46-47).

15. This was to be a replication site, with CET having been brought into the city several years ago by the mayor's office. For reasons that remain somewhat murky, but that include a concerted effort by existing community groups to "lock out" the "interloper" from the West Coast, the CET replication in Chicago never got off the ground.

6

PEER-TO-PEER NETWORKS ENGAGED IN WORKFORCE DEVELOPMENT

Peer—*n.* one that is of equal standing with another.

—*Webster's New Collegiate Dictionary* (1979, p. 838)

THE CHICAGO JOBS COUNCIL

The Chicago Jobs Council (CJC) is a peer-to-peer network of community development corporations (CDCs) and other company-based organizations (CBOs) that was created to monitor and advocate for job training and placement policies and programs. In 1981 it was formed as a collaborative between the secular Hull House Association, the Community Renewal Society (a 114-year Church of Christ endowed organization), and the Chicago League of Women Voters. The stated mission of the alliance has, from the beginning, been to "expand employment opportunities for all city residents, with an emphasis on the poor, racial minorities, the long-term unemployed, women, and others who experienced systematic exclusion from employment and career mobility." In recent years, CJC has begun to design and manage finely targeted operating programs through its many member groups.

Diversity of Membership

CJC's initial membership included the three founders and 15 other organizations. The network now has more than 90 members, most of them civic or nonprofit organizations, although the council also includes some individual members (individual members often belong to organizations that do not wish to join formally, such as the Legal Assistance Foundation and local newspapers). Approximately 70% of members are citywide organizations, with the balance based in a particular community. This distinction between citywide and community-based orientation is becoming less sharp in Chicago and probably less important. Most citywide groups are more effective if they have a base in the neighborhoods, even as neighborhood organizations are developing coalitions throughout the city to enhance their political impact. Some public agencies, such as the Illinois Department of Public Assistance, are also members. A number of reciprocal membership agreements exist, notably that between CJC and the Chicago Association of Neighborhood Development Organizations, a largely white citywide network, and that between CJC and the Chicago Workshop on Economic Development (CWED), which is mostly African American.

CJC's current board of directors includes representatives from the League of Women Voters, Women Employed, Jewish Vocational Services, the Archdiocese of Chicago, the Chicago Manufacturing Institute, the Safer Foundation, Shorebank Advisory Services, the Health and Medicine Policy Research Group, the Illinois Department of Employment Security, and faculty from the University of Chicago and the University of Illinois-Chicago's Center for Urban Economic Development.

In 1996, CJC operated with a budget of only slightly more than $400,000, which was generated primarily from membership fees and private foundation grants. This supports a director, Robert Wordlaw, and four other full-time staff persons. Case writer Marcus

Weiss (working from field notes originally collected by Gloria Cross) observes that most of the actual programmatic activity is performed by volunteers from the member organizations who are formed into lively working groups. Currently, three such groups are active.

Welfare-to-Work Group

The welfare-to-work group focuses on advocating for programs and policies that support public assistance clients in making the transition from welfare dependency to self-sufficiency. CBOs involved in the welfare-to-work group cooperate with other organizations to shape or reform welfare-to-work policy and legislation at both the state and federal levels. Members of the group produce publications, cosponsor public forums, provide testimony on welfare reform at related hearings, and distribute issue briefings to political candidates. Participating CBOs have also successfully pursued the imposition of accountability measures for statewide job training programs and influenced program (and even evaluation) design modifications for various training endeavors.

Recently, this working group launched the JOBS Case Management Training Project to design and coordinate the training of JOBS case managers in Chicago. Representatives from member CBOs staff the training teams. The training of caseworkers by organized representatives of those caseworkers' clientele has attracted considerable attention to CJC.

Health Care Working Group

Several of CJC's member groups have identified training for jobs in the health care sector as a priority. This new working group is in the process of developing a collaboration with the Chicago City College to promote such targeted training.

Workforce Development Group

The workforce development group undertakes or supports research to measure the progress of job creation activities. The group continually strives to refine a vision for more comprehensive workforce development systems. Many of the participating CBOs collaborate on influencing the leverage of governmental spending to maximize the targeting of job opportunities for low-income Chicagoans. For example, this group has collaborated with homeless service providers to identify and secure funding for creative employment approaches to serving that constituency. Advocacy with the city council is another of its important tasks, particularly with regard to monitoring compliance with the McLaughlin Ordinance, which mandates formal city inclusionary programs for firms headed by women or minorities seeking to engage in contracts on city construction and other projects.

Recently, CJC's workforce development working group obtained foundation support for a multiyear initiative, the CBO Linkage and Referral Program, funded primarily by the MacArthur Foundation and the Chicago Community Trust. The primary emphasis involves advocacy aimed at maximizing access to new jobs by low-income individuals and the creation of a shared computer database, the Chicago JOBNET. When CJC staff began to explore appropriate software for the sharing of the computerized database, board members were engaged to provide information about similar software that their respective organizations had used in their own workforce development activities. Board members such as John Plunket of the Suburban Job Link Corporation traveled to other nonprofit sites with CJC staff to evaluate customized software packages available for replication in the CBO Linkage and Referral Program. Ultimately, CJC found the appropriate software package already in use by another nonprofit organization.

Collaboration does not just happen. Over the years, CJC has developed a number of organizational mechanisms for promoting it. For example, the council established a system by which collaborating member organizations would share the "credit" for implementing a program in the records of sponsoring funders. The same is true for the city's Bank Loan Participation Program, under which Chicago provides "blended" (subsidized) loan rates to employers who agree to hire through CJC jobs networks.

Another example is afforded by CJC's role as Chicago convenor for the national demonstration program, Bridges to Work. Designed and managed by Public/Private Ventures, Inc., with funding from the U.S. Department of Housing and Urban Development and several large foundations, five programs in urban areas around the country will be engaged in operating special transportation-job linkage programs to connect low-income individuals from poor neighborhoods with jobs distributed throughout the regional economy. These connections involve transporting inner-city residents to businesses in growing airport corridors and suburbs and towns.

In most of the demonstration communities, the official metropolitan transportation organization serves as the convener of participating organizations. In Chicago, however, CJC was seen as the pivotal convener of important participants because it had a positive track record of working with the mayor's employment training personnel, the county service delivery administration, and important nonprofit and private organizations, including Suburban Jobs Link Corp., CWED, and Pace Suburban Bus Company. Observers of the Chicago workforce development institutional community continue to marvel at CJC's ability to pull these sometimes-competing organizations together and to get them to share employee referrals and job openings in an atmosphere in which scarce Job Training Partnership Act (JTPA) funding commonly fosters competition more often than cooperation.

*Organizing a "Voice" for Communities in
Confronting What Some See as an Elitist
One-Stop Career Center Initiative*

In a number of cities, including Cleveland, Pittsburgh, and (perhaps) Chicago, some see a new wave of corporate elite-driven economic development (Hill, 1996; Keating, Krumholz, & Metzger, 1995). In the context of the shift toward devolution of what had long been federal social and economic development policies and programs to the states and localities, some worry that an opening is being created for local business elites to gain nearly exclusive control over, for example, the new Workforce Development Boards.

As has been shown throughout this book, getting private firms to become serious stakeholders in urban and regional development efforts is devoutly to be wished. The Center for Employment Training (CET) and Quality Employment Through Skills Training (QUEST) cases illustrate how job training, placement, retention, and worker motivation can be greatly enhanced by company engagement in curriculum design, instruction, and overall program governance. Corporate stakeholding, however, is not the same as total corporate control over the workforce development process. In Chicago, the CJC is returning to its advocacy origins in challenging the directions that the new city Workforce Development Board appears to be taking. CJC and its members were actively involved in the popular movement 3 years ago that ultimately forced the mayor to restructure an especially inflexible and unresponsive private industry council. Into that vacuum, a citywide board is being created that has effectively no CBOs, CDCs, or other neighborhood-based groups in positions of power.

Moreover, according to CJC Director Wordlaw, not only does the new Workforce Development Board not represent Chicago's extraordinarily rich mix of CBOs but also it is dominated by the very largest corporations and institutions in the area, such as IBM, Ameritech, and the City College. There is practically no representation of small and medium-sized private firms. These are the businesses that, although accounting for many of the region's job opportunities, generally lack the resources and staff depth to be able to engage to any great extent in their own individual human resources planning. These firms could benefit enormously from participating in associations that pool resources and labor needs—including the new Workforce Development Board, from which they too have effectively been excluded.

CJC, representing communities across the city, thus finds itself in the almost paradoxical situation of advocating for the inclusion in the process of these smaller private firms. In short, because of its community network-based advocacy agenda, it urges the private sector to build more inclusive networks to govern the workforce development process in Chicago.

THE PITTSBURGH PARTNERSHIP FOR NEIGHBORHOOD DEVELOPMENT

The Pittsburgh Partnership for Neighborhood Development (PPND) is one of the oldest citywide CDC networks in the United States. Throughout the 1980s, it was the subject of attention and interest by advocates of community development, academics and scholars, and the national media (Brophy, 1993; Kotlowitz, 1991; Lunt, 1993; Metzger, 1992; Pierce & Steinbach, 1987). Partnership board members have included senior bank officers; executives of companies, foundations, and hospitals; high-level public-sector officials; university and community college deans and faculty; and such prominent community activists as the executive director of the Bidwell Training Center and founder of the Manchester Craftsmen's Guild, himself a recent recipient of a MacArthur "genius" award. As such, reports case writer John Metzger, Assistant Professor of Urban and Regional Planning at Michigan State University, PPND and its network of 10 CDCs (Table 6.1) have been deeply involved in the recent history of Pittsburgh. The communities its CDCs call home cover a major share of the area (Figure 6.1).

A Brief History of the Partnership

The Pittsburgh Partnership for Neighborhood Development was created in 1983 by the Ford Foundation, the Pittsburgh Foundation, the Howard and Vera Heinz Endowments, and the city of Pittsburgh. Grants from a number of banks, including Mellon, Pittsburgh National Corporation (PNC), Equibank, and Union National, supplemented foundation funds and the city's own Community Development Block Grant (CDBG) dollars. As an "intermediary" network, the partnership subsidizes the operating budgets of its 10 member

Table 6.1 The Pittsburgh Partnership for Neighborhood
Development Community Development Corporations

Corporation	Year Incorporated
Bloomfield-Garfield Corporation	1976
Breachmenders, Inc.	1980
East Liberty Development, Inc.	1979
Garfield Jubilee Association	1983
Hill Community Development Corporation	1987
Homewood-Brushton Revitalization and Development Corporation	1983
Manchester Citizens Corporation	1969
North Side Civic Development Council	1954
Oakland Planning and Development Corporation	1980
South Side Local Development Company	1982

groups, which are engaged in residential, commercial, and industrial real estate development and (recently) job training and placement.

As a "peer-to-peer network," the partnership CDCs have established informal and formal connections among themselves—sometimes facilitated directly by PPND's own staff—to advance their employment and industrial retention initiatives as well as other community development objectives. For the CDCs, the PPND network expanded the pool of private funds for their work, especially in real estate; provided a common forum for problem solving, agenda setting, and technical assistance; supplied a steady source of operating capital in an era of federal spending cutbacks; and simplified the time-consuming task of fund-raising.

The involvement of the philanthropies was particularly nurtured by the Ford Foundation, which viewed the partnership as an experiment and subsequently as a model for a collaborative that could establish a local CDC network. Ford's Urban Poverty Program solicited the involvement of the other philanthropic organizations (as it would later do in other cities) to establish public-private funding partnerships that would catalyze and support the expansion of local CDC networks.[1] The large commercial banks in Pittsburgh viewed the partnership as an opportunity to broaden their involvement with CDCs beyond the traditional participation of branch managers on individual local CDC boards. Before the Community Reinvestment Act (CRA)

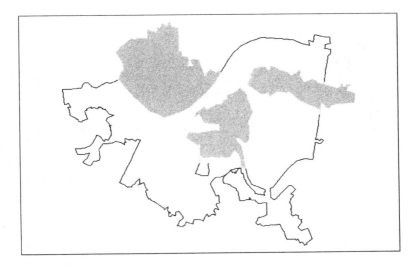

Figure 6.1. Map of Pittsburgh Neighborhoods Served by Partnership CDCS (Shaded Areas)
SOURCE: Reproduced With Permission From J. Metzger

became a crucial factor in stimulating banks' interests in neighborhood development, Pittsburgh banking leaders, such as Edward Randall (PNC) and Donald Titzel (Mellon), lent their influence to aiding PPND's access to other corporate leaders' specialized technical expertise and public and private sources of additional funding.

PPND's first full-time executive director, Sandra Phillips, left the organization in the winter of 1996 and 1997. Before joining the partnership in 1988, she had been a prominent, imaginative, and outspoken CDC director in the Oakland neighborhood of Pittsburgh. At that time, she was named to the Federal Reserve Bank's National Consumer Advisory Council and chaired its CRA committee when pivotal changes to CRA regulations and enforcement were being implemented. She also served as director of the Pittsburgh branch of the Federal Reserve Bank of Cleveland.[2]

As a result of CRA regulatory revisions in 1995, organizations such as PPND became eligible for bank funding as "community development investments." The partnership represented an important vehicle for Pittsburgh banks to collectively and individually engage in mutually supported endeavors. For the banks, the partnership has identified new business opportunities, provided a means by which the banks can collectively manage and share risk, improved their CRA performance, enhanced their own civic stature, and demonstrated that support of community development does not necessarily slow the career advancement of participating bank executives—a well-known prejudice in the field.

PPND also functions as an "intermediary network" that makes grants and loans to CDCs to contribute to their annual operating costs and the financing of their real estate projects. The "core" budget expenses of the CDCs and two citywide technical assistance agencies are supported by partnership funds. PPND also makes special grants to increase the capacity of CDCs to meet identified community needs through staff training, planning and economic development activities, and programs that develop tenant leadership in government-subsidized housing.

Real estate investments by the partnership attract additional public and private funds into CDC projects. According to an internal audit, during 1993 every PPND dollar was claimed to have leveraged approximately $5 for commercial deals and $15 in the case of housing projects. Each CDC and community organization supported by partnership operating funds has also secured project financing from PPND. In addition, the partnership has helped to finance real estate developments undertaken by other community-based nonprofits not directly backed by PPND operating money.

By the end of 1995, partnership CDCs had built 817 units of low- and moderate-income housing: 55% for home ownership and 45% for rental. Through 1993, a majority of the homeowners in partnership projects were minorities and women. Among tenants, nearly all were minorities and more than two thirds were from the targeted neighborhood. These CDCs have also developed 640,000 square feet of commercial and industrial space. The partnership estimates that through 1993, investment from its own Development Fund and from Ford's Program Related Investment window have created 424 jobs.

PPND as a Network for Employment
Training and Workforce Development

Some of the CDCs funded by the partnership for real estate development have also created their own "hub-spoke" networks for job training and placement, either through the direct financial sponsorship of PPND or as a city-funded "Neighborhood Employment Project." Since 1987, the municipal government has combined its federal JTPA funds with CDBG money to support local employment centers in Oakland, the East End, and the east, north, south, and west sides of Pittsburgh, some of which are controlled by or involve partnership CDCs (Deitrick & Harrison, 1994). These networks are designed to connect with employers, training vendors, and other community organizations in CDC service areas as well as to outside public and private institutions operating citywide or in the region. Through the city's JTPA- and CDBG-funded Neighborhood Employment Program (NEP), CDCs assist residents in locating jobs and acquiring needed skills by networking with local employers and education, training, and placement providers. The CDCs identify residents and typically offer guidance, counseling, and postplacement monitoring services.

One of the most prominent of these hub-spoke networks was the Job Links program, affiliated with the Oakland Planning and Development Corporation (OPDC), an early member of the partnership (Higdon, 1993). Job Links originated during the 1980s when the Oakland business district experienced a surge in employment and construction activity related to the expansion of the universities, hospital, and institutions headquartered in the neighborhood (Weiss & Metzger, 1987).

Job Links is composed of four elements. Employment readiness training emphasizes basic job search and workplace skills. Job development outreach builds relationships with employers and identifies labor force needs. Job placement assistance links residents with participating local employers. Ongoing monitoring and tracking are conducted to facilitate counseling, training referrals, and program evaluation. The job readiness training consists of a 3-week course offered 10 times each year, with an annual enrollment of approximately 100 persons drawn from neighborhood contacts, organized promotion, and by word-of-mouth. The program is targeted to the unemployed and underemployed, many of whom lack work experience or needed skills. By 1993, more than 80% of those in the program were African Americans, two thirds were at most high school graduates, and many were single mothers.

The Job Links staff includes a full-time community counselor who screens candidates for training and jobs and a part-time marketing coordinator specializing in job development, both of whom are employed by OPDC; a job readiness trainer employed by another small neighborhood CDC named Breachmenders; and the executive directors of OPDC and Breachmenders, who devote a portion of their time to program administration, fund-raising, and outreach. The two CDCs manage the program as a joint venture. The operating budget of Job Links is now directly funded by the partnership. Other sources of financial support in recent years include the Pittsburgh Foundation; SEEDCO (a Ford Foundation spin-off based in New York); various corporate, religious, and bank-affiliated philanthropies; the Pennsylvania Department of Community Affairs; and federal CDBG monies set aside for allocation by Pittsburgh City Council members.

For a variety of reasons, the city agency that administers the federal JTPA funded an alternative employment initiative based in Oakland known as the Joint Oakland Based Neighborhood Employment Team (JOBNET). According to an evaluation conducted at the University of Pittsburgh (Higdon, 1993), this program, which is located at the Carnegie Institute, serves a somewhat better educated, higher-income, and more geographically dispersed clientele than does Job Links and operates as part of the citywide network of JTPA-funded NEPs. For the city, JOBNET offers the attraction of engaging a major cultural institution in the regional employment training system.[3]

Job Links has forged strong ties with the University of Pittsburgh and its affiliated Medical Center, which are the dominant Oakland area employers, as well as with other local businesses that have long-standing relationships with OPDC and Breachmenders. In particular, financial institutions that work with these CDCs through the partnership and the Pittsburgh Community Reinvestment Group have hired Job Links referrals to work in neighborhood branches and other corporate locations. In addition, the Hill CDC sends graduates of its Hill Employment Linkage Program (HELP) construction training to the Job Links job readiness workshops. PPND has used Job Links as a learning model in its efforts to build the employment program capacity of some of the weaker CDC-affiliated NEPs.

Over time, the role of the partnership has evolved to become the direct funder of the Oakland Job Links operating budget as well as the partial staff costs and overhead devoted to employment training by the other partnership CDCs. Where JTPA is involved, PPND in effect pays for the "soft" expenses of outreach, job development, counseling, and administrative compliance that are incurred by those CDCs affiliated with NEPs but not otherwise reimbursed by the government. PPND estimates that, during 1995, the JTPA Neighborhood Employment Centers and the CDC programs such as Oakland Job Links, HELP, which is operated by the Hill CDC, and various youth-targeted efforts (not including Youthbuild) had made 566 placements into full-time and part-time positions.

The partnership would like to establish Job Links as a central intake vehicle for all the CDCs, which would identify job openings among local institutions, manufacturing firms, and neighborhood businesses and refer residents to the employment readiness program (Figure 6.2). This model would include a formalized linkage with the highly regarded Bidwell Training Center, whose director, Bill Strickland, is now a member of the PPND board, and would emphasize the creation of opportunities for low-wage job experience for those who lack work histories before referrals are made to employers. Although this would indeed represent a considerable integration of actors and institutions throughout the city, the recent withdrawal of financial support by some of the key funders makes it unlikely that such a comprehensive network will be implemented any time soon.

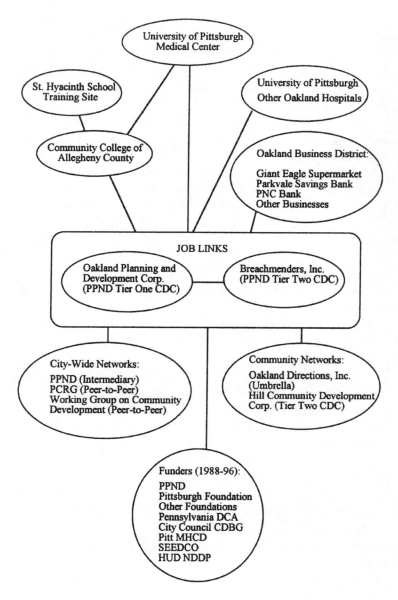

Figure 6.2. Oakland Job Links Hub-Spoke Network
SOURCE: Reproduced With Permission From J. Metzger

Pittsburgh Manufacturing and
Community Development Network

With the assistance of Robert Gleeson, a board member of the partnership and also coauthor of an influential white paper on the regional economy sponsored by the Allegheny Conference on Community Development in 1993, PPND decided that interfirm networking facilitated by CDCs could become a viable strategy for retaining and strengthening manufacturing industries, and that this could result in expanded job opportunities for neighborhood residents. Existing state and local economic development programs were not pursuing this approach, which views industrial outreach as a form of community organizing (recall the description of the Austin Initiative on the West Side of Chicago).

Since 1994, the partnership has been cultivating a peer-to-peer network among CDCs actively engaged in industrial retention and the manufacturing companies in their service areas. This strategy recognizes the importance of network building among large and small firms in sustaining local and regional economic development as well as the vital role of manufacturing employment and industrial districts in the economic revitalization of urban neighborhoods. For the partnership, the peer-to-peer network of manufacturers and CDCs holds the promise of expanding the realm of PPND influence in economic development, strengthening the ties between CDCs and local manufacturing firms, and enhancing the industrial base of Pittsburgh through networked linkages between competing businesses, their suppliers and customers, and CBOs. This networking can then improve the management systems of the participating firms, the overall competitiveness and economic performance of local industrial districts, and the employability and incomes of neighborhood residents (Rosenfeld, 1995a).

In 1993, the partnership funded a pilot project involving two of its CDCs, the Homewood-Brushton Revitalization and Development Corporation and East Liberty Development, Inc. East Liberty and Homewood-Brushton are adjacent neighborhoods that serve as home to one of the important remaining industrial centers within

Pittsburgh. A private company, World-Class Industrial Network (WIN), was hired to design and manage the network. WIN is headed by Barry Maciak, formerly an industrial retention specialist with the Southwestern Pennsylvania Industrial Resource Center (SPIRC) and currently the executive director of the Connelly Center for Entrepreneurship at the Duquesne University School of Business. At SPIRC, Maciak's job had been to make contacts with CDCs and manufacturing companies.

With technical support from graduate students at Carnegie Mellon University and a grant from a local philanthropic foundation, the first task of the network was the compilation of a computerized and mapped inventory of 64 establishments located within the geographic boundaries served by Homewood-Brushton Revitalization and Development Corporation (HBRDC) and East Liberty Development, Inc. (ELDI). The local knowledge possessed by the CDCs opened doors to firms that were never reached by SPIRC, which has a mostly suburban focus. WIN then positioned itself as a resource for the CDCs, providing information and guidance on issues related to manufacturing, business management, industrial retention, and economic development. Six manufacturers located in East Liberty and Homewood were initially recruited to become the steering committee of the network. This group included Nabisco Foods and Matthews International, the two largest employers in the area, as well as Hoechstetter Printing and three small supplier firms. Provided they employed at least 20 workers, additional companies were added through contacts made in the community networks of the CDCs and the outreach and questionnaire surveys of SPIRC and WIN.

By the end of 1994, eight companies had joined the original six, two other CDCs supported by the partnership (the Lawrenceville Development Corporation and North Side Civic Development Council) became involved in the network, and the manufacturing inventory expanded to include the industrial districts of Lawrenceville and the north side. Four firms from Lawrenceville have since been added to the network, which now encompasses a citywide industrial corridor stretching from Homewood and East Liberty on the east side to Lawrenceville and across the Allegheny

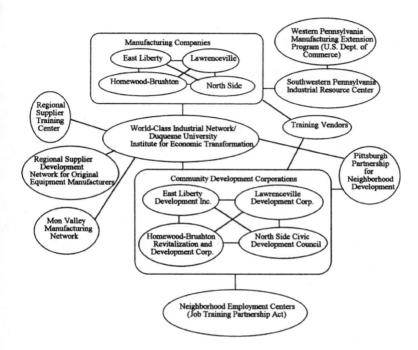

Figure 6.3. Pittsburgh Manufacturing and Community Development Peer-to-Peer Network
SOURCE: Reproduced With Permission From J. Metzger

River to the north side. The categories of organizations that currently make up the network are depicted in Figure 6.3.

For the smaller companies, the manufacturing and community development network provides access to best-practice training and management methods that might otherwise not be readily available to them. In addition, WIN can connect these manufacturers with a supplier network program sponsored by the Connelly Center at Duquesne. The participation of the small firms is constrained by their limited time and resources (some of the small printing firms have in fact already dropped out, confronting their own crises of ownership transition or declining sales). This is less of a concern for the larger companies, which possess more in-house resources and capacity to donate. It is hoped that the pressure on these "leaders" will diminish as the network grows and achieves a broader base of company and employee participation.

The manufacturing and community development network was explicitly designed as an industry-led effort that would first produce benefits for the participating companies and then later result in new employment linkages with the targeted communities. The partnership CDCs have been meeting on their own and apart from the manufacturers in this first phase to identify and refer companies to the network, conduct industrial surveys and outreach, develop their employment training capacity, assess the skills and needs of local residents, and establish a communication system among themselves and with the network companies. The initial interviews conducted with the companies recruited to the network have produced specific and immediate intentions by the CDCs and SPIRC, such as packaging financing for equipment purchases, locating possible plant expansion sites, informing the neighborhood employment projects of factory job openings, and making management training referrals to the Connelly Center at Duquesne.

The separate activities of the companies and CDCs in the network are already converging on the topic of employment training. Some of the network firms already employ neighborhood residents, but most have made little or no connection to the partnership CDCs on this issue. The community development corporations are not perceived by the network manufacturers as reliable sources of employment referrals. Nabisco, for example, depends on the East Liberty office of the state Job Service, not on PPND or its member CDCs, to provide it with craftworker referrals.

Staff turnover and a history of somewhat superficial contact with these firms has heretofore limited the capacity of the CDCs to broker job training and placement services for the network. For example, the two founding network CDCs, HBRDC and ELDI, each changed executive directors twice during 1994 and 1995 and were both being managed by the same interim consultant while permanent replacements were being recruited.

Another missing connection is with the Community College of Allegheny County (CCAC), an important regional provider of education and training services with a branch campus located in nearby

Homewood. CCAC in Homewood is recognized for its job readiness programs, but its occupational training mostly emphasizes the building trades (for YouthBuild workers) and computer-related and medical technician skills. The network companies, including the large plants, are demanding practical, user-oriented training approaches that are keyed to their own shop floor needs—in food processing and printing.

Problems in the Original Network of CDCs

We have noted how some of the leading partnership CDCs have been struggling in recent years. In addition, several high-profile CDC real estate projects in neighborhood commercial centers and industrial districts have suffered acute financial losses or failed due to a combination of weak economic demand and poor business and project management. Although gifts to the Partnership Operations Fund have risen, outside gifts to the Development Fund dropped substantially in recent years.

These problems are not entirely surprising, given the inherent risks in undertaking community economic development in older central cities such as Pittsburgh, a place that continues to lose jobs and population. Even the successfully performing CDC projects, especially in owner-occupied housing, however, continue to require deep subsidies from the Urban Redevelopment Authority and other public agencies. A well-funded effort to revitalize an industrial-strength bakery failed despite strong regionwide support and national media attention. Ventures that focus on commercial district revitalization have always had a mixed record and troubled history in Pittsburgh; this has only gotten worse in recent years.

High staff turnover in CDCs exacerbates the situation. The decline of HBRDC since the departure of Mulugetta Birru in 1992 was, in retrospect, an early marker of trouble ahead. Enrollment in neighborhood training programs dropped at HBRDC, staff resigned, CCAC reduced its budget for the Homewood campus, and the city folded the JTPA funds into the East End Neighborhood Employment Program, which serves a larger geographic area that includes the neighborhoods targeted by East Liberty Development. Some of the efforts of HBRDC are now being supplanted by YouthBuild Pittsburgh, which started in 1992 with federal funding from the Department of Housing and Urban Development and later from the AmeriCorps program.

By 1994, few of the initial CDC directors remained in their original positions (and Phillips has since departed). The high turnover has forced the diversion of operating funds to cover property management deficits, some of which are in buildings financed by PPND. This in turn has required putting greater staff time and energy into disclosure and reporting of the multiple sources and uses of funds in all CDC-managed properties to distinguish between deficits and simple cash-flow problems. Computerized accounting methods are finally being introduced throughout the network.

Many partnership groups have effectively abandoned trying to recruit from outside the city, turning instead to promoting new directors from within their own ranks. More money than ever before is being dedicated to staff development. These staff development and computerized accounting reforms are promising, and long-overdue reforms within PPND and its member CDCs have been undertaken in response to what is now a widely acknowledged crisis of leadership.

The Pittsburgh partnership as an organization is clearly in a period of crisis, especially regarding its desire to constitute an effective workforce development network. A recent (confidential) outside evaluation, conducted by Renee Berger Associates for the Heinz Endowments, was apparently quite critical. The ideological climate has also changed. For example, under its new executive director, Heinz (the foundation, not the manufacturing company!) appears to be rejecting (or at least demoting in importance) the former goal of community building in favor of a less "place-based" strategy whose insignia is the creation of "self-sustaining households." This is the same theme that is being used by many policymakers to rationalize the federal government's new and—in the opinion of some—regressive time-delimited welfare "reform." An influential adviser to Heinz, David Rusk, is also on record as opposing place-based "ghetto development" policies (Rusk, 1993). In that context, Heinz has substantially cut back funding for PPND.

Moreover, relations between the partnership and the city government have now become openly hostile. Although there are many local, parochial, as well as principled reasons for this, from a distance it is a most unfortunate and even paradoxical development. It is "paradoxical" because the current city government is mainly composed of former community activists and organizers, including the mayor, deputy mayor for policy, and the heads of the Housing and Urban Redevelopment agencies.

Notable about the continued successes of CET, New Community Corporation, CJC, and QUEST (among others of our cases) is their underlying financial independence from any one particular funding source, their roots in community or racial and ethnic organizing, and their ability and willingness to at least occasionally exercise independent political "voice" in support of their programmatic objectives. Individual CDCs in Pittsburgh, especially (for a time) OPDC, shared such a history. If PPND ultimately fails, however—and that is by no means a foregone conclusion—it may well be because it was too captive of the elite growth coalition that formed it—too studiedly apolitical. Only time will tell.

BUSINESS OUTREACH CENTERS OF NEW YORK CITY AND BRIDGEPORT, CONNECTICUT: AN INTERRACIAL, INTERETHNIC, AND PEER-TO-PEER SMALL BUSINESS DEVELOPMENT NETWORK

Founded in Boro Park, Brooklyn, New York, in 1990, the Business Outreach Center network (known locally as BOC) acts as a broker linking technical assistance providers—banks, management consultants, accountants, lawyers, export-import specialists, and government procurement officers—to the owners of small businesses in lower-income neighborhoods in New York City and Bridgeport, Connecticut. Our case was written by David Sweeny, who operates a shared-manufacturing business incubator in the Greenpoint section of Brooklyn that has received national attention for its approach to fostering growth and networking among tenant wood product manufacturers.[4]

The case documents the evolution of the BOC system from a "good idea" implemented in one particular neighborhood, involving one particular ethnic group, into a fairly centralized network of such small business development brokers operating throughout the region and involving a remarkably racially and ethnically diverse mix of members. Recently, the individual neighborhood-based nodes in the network—the local BOCs—have begun acquiring greater autonomy. They are also focusing more on issues of employment training and workforce development.

Each of the eight BOCs is sponsored by a successful, neighborhood-based human services or community development organization (in New York City, they are called "local development corporations" [LDCs]), with years of experience in the field of small business development (Figures 6.4 and 6.5). As the BOCs successfully brokered resources from banks, leasing agents, business planners, marketing consultants, and venture assistance programs within universities and government agencies for their small business owners situated in their relatively isolated neighborhoods, they have achieved acclaim in the media and among political leaders as a vehicle for integrating local small business growth and community development (Figure 6.6) (Kennedy, 1994).

The collaboration among the eight (to date) BOCs in the network sustains and enhances their individual capabilities in their target communities. A 1994 survey of 74 randomly selected clients indicated that, in that year, the network served approximately 350 businesses, many of which were minority or women owned ("BOC Network Service," 1995). Collectively, the BOCs use information exchange, biweekly interaction, and cooperative referrals of clients among one another. They share centralized management, which is based at the Boro Park headquarters. The centralized staff assumes the burdens of program evaluation, enforcing service quality control, public relations, and fund-raising. Individually, the BOCs are free to concentrate on assessing the quality and results of referral contracts between service providers and disadvantaged businesses. Each BOC has its own advisory board, typically composed of lawyers, accountants, bankers, entrepreneurs, organization leaders, and other resource vendors.

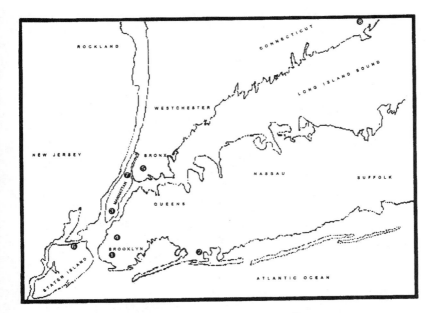

❶ BORO PARK
5224 13th Avenue
Brooklyn, NY

❷ ROCKAWAY
1931 Mott Avenue
Far Rockaway, NY

❸ CHINATOWN/LOWER
EAST SIDE
125 Canal Street
New York, NY

❹ FLATBUSH
855 Flatbush Avenue
Brooklyn, NY

❺ HUNTS' POINT
961 Southern Blvd.
Bronx, NY

❻ STATEN ISLAND
1207-09 Castleton Avenue
Staten Island, NY

❼ HARLEM
102 W. 116th Street
New York, NY

❽ BRIDGEPORT
Bridgeport Economic
Development Corp
10 Middle Street
Bridgeport, CT

❾ BOC NETWORK
EXECUTIVE OFFICE
5224 13th Avenue
Brooklyn, NY

Figure 6.4. The Business Outreach Center Network as of Spring 1996

A number of the LDCs that sponsor BOCs have for years been engaged in employment training and workforce development activity, principally with JTPA funding. Recently, both the BOC leadership in Brooklyn and the individual sites have begun to put greater effort into building on those connections to identify and address the skill training and monitoring needs of their small business clients.

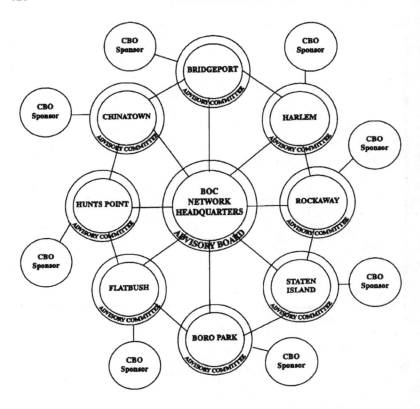

Figure 6.5. BOC Network

Origins of the BOC

The BOC was initially conceptualized by senior staff of the Council of Jewish Organizations (COJO), a multiservice agency specializing in providing employment training and diverse social services in the Boro Park neighborhood of Brooklyn. Unlike other Brooklyn neighborhoods, Boro Park's Orthodox Jewish community was not being served by an LDC. Because of Boro Park's isolation, emerging businesses in the community were underserved by conventional financial institutions and governmental-funded business assistance and capitalization programs.

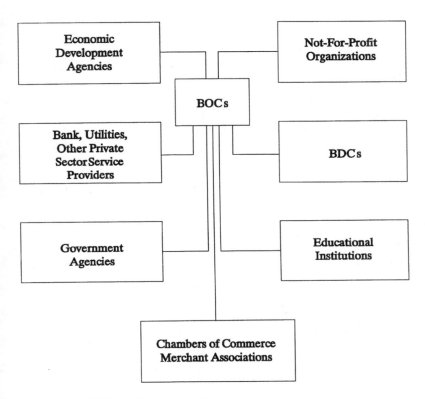

Figure 6.6. BOC and Its Linkage Partners

COJO itself was already well connected to political leaders and public agency funders (speaking at COJO's recent annual fund-raising banquet, New York's mayor noted that it was the first time he had ever seen all of his department heads in the same room at the same time!). The planners of the BOC recognized, however, that it would be prudent to avoid duplicating what other Brooklyn organizations saw as their turf. They also decided to avoid burdening COJO's social service-oriented staff with responsibilities for assisting business clients in skill areas not readily available from in-house personnel. Therefore, the designers of the BOC sought to assist local business persons in accessing valuable networks and institutions outside of Boro Park. At first, they did not consider creating an areawide network as such: That came later.

It was the BOC's initial state funding source, the New York Empire Development Corporation (then called the New York State Urban Development Corporation) that strongly encouraged COJO to seek out partners from beyond their neighborhood to demonstrate an expanded regional impact. Initially, the COJO BOC contracted with a similar organization in Far Rockaway that was led by a rabbi known to COJO leadership to be interested in comparable endeavors. The Rockaway Development Revitalization Corporation became the first satellite or licensee of the BOC network. Eventually, the network grew to include sister organizations in Chinatown, Harlem, Flatbush, Hunts Point/South Bronx, Staten Island, and Bridgeport. Collectively, the BOCs currently serve small business clients from Orthodox Jewish, Latin Hispanic, African American, Asian American, Caribbean American, newly immigrated Russian, and still other backgrounds.

The Challenge of Building a Peer-to-Peer
Network While Ensuring Quality Control

With the establishment of the additional BOC sites, it became necessary for the leadership to find ways to reconcile local responsibility with some degree of centralized quality control—a problem endemic to all such networks (CET, for example, continues to wrestle with this difficulty). A detailed, formally structured set of operating prescriptions and processes was developed by the Boro Park staff, chief among them Paul Chernick, Nancy Carin, and John Fratianni. With multiple offices now in place, Fratianni and Carin took on administrative, managerial, and coordination responsibilities within the network. Beyond simple management and coordination, BOC headquarters' staff would also take on the responsibilities of funding, reporting, and advocacy. These processes and relationships continue to comprise the basic BOC organizational structure.

Although each site creates its own network of business service providers, and has its own complement of local advisers, all are also woven into an umbrella network wherein each shares business development resources, exchanges advice about how to handle troubled business cases, and joins forces for strategic planning, fund-raising, and network operations.

Although no one pretends that interracial, interethnic collaboration is easy, these peer-to-peer alliances have promoted a degree of mutual cultural appreciation among BOC participants and their client businesses. Latino businesses now work with Jewish community leaders and Rockaway retail merchants. Each BOC office brings a unique set of opportunities to the central network, creating an unusual synergy. Moreover, the BOC network's expansion has clearly broadened its popular political base—for example, with the city and state governments—considerably improving funding prospects for the network.

The origin of the Harlem BOC is emblematic of the possibilities inherent in interorganizational networking. In midsummer of 1994, Mayor Rudolph Giuliani announced his support for the "regular" merchants along 125th Street in Upper Manhattan and agreed to relocate the many mostly African immigrant unlicensed street vendors who had set up on the sidewalks.[5] Business Commissioner Rudy Washington identified a plot of city-owned land at 116th Street and Lenox Avenue, across the street from the Masjid Malcolm Shabazz. This had been the mosque of the late Malcolm X and remains a well-known, busy Harlem street-corner gathering place. The mostly Muslim vendors looked to the mosque's imam for guidance on the relocation. Imam Pasha agreed to "adopt" the market, and the city paid for upgrading the site's infrastructure.

When the imam approached then-lieutenant governor Stanley Lundine for economic assistance, Lundine introduced the imam to the leaders of the BOC in Brooklyn. Remarkably, a strategic alliance was formed, the Harlem BOC was created, and the network now advises the Muslim vendors in Harlem on management, advertising, and uses of on-line telecommunication. They also hold twice-weekly entrepreneurial training seminars. Leaders travel regularly between the boroughs.

With local and federal funds secured, the BOC central staff designed a Request for Proposals process and, eventually, a BOC (operations) handbook. The handbook provides programmatic definitions, start-up and management prescriptions, commonly used documents and reporting forms, training techniques, and even office siting guidelines. At the same time, however, organizations proposing to establish BOCs are advised that the network antici- pates that each community will structure its operational priorities and techniques based on its own determination of local needs. Performance accountability is required by the peer-to-peer network of BOCs; evaluations tend to focus on quality of operations and on the enhancement of networking capacity rather than on quantified measures based on intake numbers or counts of "successful" ven- tures or job placements.

Linkages between the BOC network and specific industry-related organizations in the apparel trades have resulted in several cross- borough connections. In one instance, an emerging entrepreneur in Brooklyn was referred to the Garment Industry Development Corporation (GIDC) in Manhattan. The BOC facilitated the entre- preneur's obtaining access regarding technical sewing training and leads for contract work. Most important, trainees were also referred to the Brooklyn firm by the GIDC. In another instance, a Brooklyn garment manufacturer was referred by the Boro Park BOC to the Chinatown BOC. Staff in Chinatown connected the entrepreneur to unions and other garment industry organizations that played a significant role in the needle trade. In other cross-ethnic examples, a Filipino-headed firm from Flatbush was similarly connected to Chinatown garment manufacturers as a result of cooperation be- tween the Flatbush BOC and the Chinatown BOC. Similar services were provided for a Russian entrepreneur in Brooklyn who sought access to Chinese silk-screening subcontractors operating in China- town. In a venture unrelated to the garment industry, the Chinatown BOC assisted a client of the Borough Park BOC who manufactured vertical blinds and was connected to the Taiwanese trade associa- tion for leads regarding equipment purchases in the Far East. These examples (and numerous others) relate to the cross-community and cross-ethnic cooperation that regularly find different BOCs within the network connecting client entrepreneurs to subcontractors, trade associations, unions, worker referral programs, and export assistance providers.

Establishing a More Independent, Secular Identity

The evolution of the BOC system underwent a major step when, early in 1996, the network severed official legal ties with the Council of Jewish Organizations of Boro Park and incorporated as an independent entity (although central management personnel remain at that location in a space donated by COJO). This transition is not so much a "break" away from COJO as it is the result of a realization that the network would be strengthened by establishing a level of autonomy, mutual support, and self-sufficiency. At the same time, within each member's neighborhood, the BOCs are making themselves increasingly independent of their original "parent" LDCs.

This stage of organizational evolution has not been without stresses and strains. Partly to avoid charges by federal grant makers of "double-dipping" by the LDCs, their associated BOCs are being required by the leadership to establish separate office locations. Regulations about the nonmingling of funds and personnel are reinforced through meetings and site visits from central management staff and directors of other BOCs. Sweeny notes that interference, "noncompliance," or other departures from the mutually adopted BOC network approach are addressed by gentle prodding, friendly arm twisting, and (on rare occasions) subtle suggestions of "expulsion" and defunding. The challenge for the network will be to sustain such subtle and basically nurturing forms of quality control while further shifting from central management by COJO to a board of directors chosen by the member sites.

Workforce Development Implications

Although its activities emphasize business assistance, jobs and training are becoming an increasingly important organizational objective of the BOC network. The central BOC staff have begun to collect and consolidate job development information from each of the sites. Core BOC staff are regularly invited to participate in high-level planning and training meetings with the city's JTPA program administrators and trainers. Whenever possible, BOC staff attempt to link their emerging entrepreneur clients with workforce development programs in their own neighborhoods and, occasionally, in neighborhoods served by other BOCs.

Despite the autonomy sought by BOCs from their sponsoring LDCs, they remain aware of their important role of providing a linkage between "parent" entity training programs and the growing firms assisted by the BOCs. Thus, staff share information with local training program personnel about activities and job trends among assisted firms and regularly make presentations to the LDCs' job training classes and related workshops. Whether the BOCs' operating emphasis on achieving the trust of local companies, by offering free services and a business-oriented delivery approach, will also meet the needs of neighborhood residents for good jobs at good wages (or at least for attractive opportunities to enter the world of work with a reasonable chance of future mobility) remains to be seen.

NOTES

1. For a recent evaluation of five of these Ford-sponsored citywide collaboratives, see Nye and Glickman (1995). The Rutgers group is currently examining 17 such city partnerships; see Glickman and Nye (1996).

2. Another CDC director, Mulugetta Birru, succeeded Phillips on the Federal Reserve's National Advisory Council and eventually became the director of the city of Pittsburgh's Urban Redevelopment Authority.

3. Pittsburgh JTPA administrators and OPDC have been unable to reach agreement on the possible inclusion of Job Links in the city-funded NEP system. The dispute has centered on JTPA regulatory requirements (a common source of tension for CDCs throughout the country that work with JTPA) and the desire of OPDC to remain independent from city hall in negotiations with local employers. The city-funded JTPA NEP system, however, which allocates CDBG funds to CDCs above and beyond what is distributed through PPND, has remained outside of the control of the foundation-dominated PPND. The closeness or distance between CDCs and city governments varies tremendously throughout the country, even within our small sample of case studies. In other words, the long-standing tension between PPND and the city is by no means unique.

4. For an excellent introduction to the Greenpoint Manufacturing and Design Center, see Isabel Hill's (1995) video.

5. This account is taken from Wylde (1996, p. 132).

REGIONAL INTERMEDIARIES BRIDGING BUSINESS DEVELOPMENT, COMMUNITY BUILDING, AND JOB TRAINING

If the community economic development and school-to-work movements are to advance, they will need to find ways to work together to bridge racial and geographic barriers, draw people into long-term interactions and development strategies—and overcome the isolation and lack of connectedness to the economy that is the guarantee of continued poverty.

—Austin (as quoted in *Jobs for the Future*, 1995, p. 1)

MULTISTATE PUBLIC AND PRIVATE NETWORKS AS A LOCAL DEVELOPMENT VEHICLE: THE REGIONAL ALLIANCE FOR SMALL CONTRACTORS

This is a story about a different kind of training—entrepreneurial training—that again uses the organizing principle of operating through boundary-spanning networks. The Regional Alliance for

Small Contractors initially emerged in 1989 from an initiative within the Port Authority of New York and New Jersey and its Office of Business and Job Opportunity. The Greater New York and Northern New Jersey region and the Port Authority were planning more than $50 billion in capital improvement expenditures in the ensuing 5 to 8 years, with a mandate to ensure that not less than 10% of related construction work would be contracted to minority firms (a number of federal programs with respect to transportation and defense allocations contain such 10% minority business inclusionary goals). The innovators of the alliance concluded that previous minority inclusionary programs were inadequate and that achieving sustained success for emerging minority firms would require a more sophisticated series of relationships among prime contractors, financing and bonding institutions, and with key operatives responsible for managing massive infrastructure investments by city, state, and regional agencies.

The alliance endeavored to overcome the lack of access of small, minority-, and women-owned construction firms to critical, industry-specific business skills with mentoring, financing sources, timely bid information, and networking relationships. Our case writer, Leslie Winter, an expert on community development and real estate finance and former director of the Office of Real Estate for the New York City Office of Business Development, depicts the key programmatic components of the alliance as follows:

- ◆ Managing growth: a structured series of educational and such technical skill courses as bid estimating
- ◆ Loaned Executive Assistance Program: providing pro bono highly experienced executive mentors from among the largest firms in the construction industry
- ◆ Opportunities Marketplace: an accessible and affordable weekly information source and report on construction contracting opportunities
- ◆ Financing Small Contractors (FISC): a credit enhancement and performance program linking small construction firms to senior banking and surety firm executives

These activities are pursued in terms of the following four programmatic objectives:

- Increase the number and size of contracts awarded to small, minority-, and women-owned businesses (S/M/WBEs) and expand the range of their business opportunities
- Expand the capacity of S/M/WBEs to undertake contracts of increasing size and complexity
- Strengthen the regional construction industry through enhanced production contributed by an increasingly competitive pool of S/M/WBEs
- Promote the representation of minorities and women in the ownership, management, and workforce of participating companies without excluding men and nonminorities

Making Top Corporate Executives Stakeholders

From the earliest stages, the preliminary planning teams for these initiatives included leaders from nationally prominent construction contractors and construction management firms, such as the Bechtel Corporation, accounting executives from Deloitte & Touche, and several public agency officials who controlled billion-dollar expenditures. Within a short period of the commencement of planning roundtables for the endeavor, executives from other major firms were contacting Port Authority staff and volunteering to be partners in the proposed alliance initiative. Within a year of exploring the concept, Port Authority staff had secured $600,000 from the authority and other contributing state and local agencies in start-up funding and 31 prominent partners from the private and public sectors.

The steering committee understood the value of insisting that only the chief executive officers (CEOs) and commissioners would sit on the board of directors of the newly incorporated alliance. This participation of CEOs and agency heads extended beyond annual meeting attendance to active networking and support during alliance-sponsored banquets and even discrete lunches with key federal officials and insurance firm executives in Wall Street private dining clubs. Such activity extended to participation as cochairs (one public sector and one private sector) of a series of program committees that met quarterly. Figure 7.1 displays the corporate membership of the alliance as of spring 1996; this is a truly impressive display. Also, it has paid off; internal audits show that more than 1,000 minority- and women-owned firms (plus an additional 200

Sustaining Partners
Chemical Bank
HRH Construction Corp.
Morse Diesel International, Inc.
O'Brien-Kreitzberg & Assoc.
Tishman Realty & Construction Co., Inc.
Turner Construction Company

Associate Partners

AJ Contracting Company, Inc.	Lehrer McGovern Bovis, Inc.
Baer Marks & Upham	Public Service Electric & Gas Company
Bechtel Corporation	Sierra Mechanical Corporation
Integral Construction Corp.	Zwicker Electric Co., Inc.

Partners

Andrew & Arthur, Ltd.	Forest City Ratner
Asbestos Carting Corporation	L. K. Comstock & Company, Inc.
Barney Construction	Livel Mechanical & Equipment Corp.
Bradford Construction	Nelson Maintenance Services, Inc.
Crow Construction Company	Raytheon Constructors Int'l
CSR Construction Corp.	Small's Mechanical Contractors, Inc.
Don Todd Associates	TAK Construction Inc.
Emcor Group, Inc. (formerly JWP Inc.)	TAM Construction Corp.
Evanbow Construction	York Hunter, Inc.

Supporters

Atlantic Electric Company	Darryl E. Green & Assoc., Inc.	Mitchell/Titus & Company
Basser Construction	Deloitte & Touche	Parsons Brinckerhoff, Inc.
Cayuga Construction	Frederic R. Harris, Inc.	Watson, Rice & Company
CRSS	Impressions Construction, Inc.	

Small Business Supporters

A & T Iron Works	Gibralter Waterproofing & Rest.	Prospect Electric Service, Inc.
Achilles Construction, Co., Inc.	Hispanic Ventures	QDR Develop. Corp.
Ampere Electric, Inc.	James F. Volpe Electric Constr. Corp.	Quantum Constr. Corp.
Armand Corporation	JCS Contracting & Development Corp.	Radcliffe Decorating
Becom Real, Inc.	Keith Robinson Construction, Inc.	Rich Associates, Inc.
Carter's Construction	Kevco Electric, Inc.	Riverside of NY Constr., Inc.
Chas. James Pipe. & Heat., Inc.	L/O Inc.	RMKG, Inc. d/b/a/ RG Assoc.
Clawill Plumbing & Heating	Mt. Etna Electric Corp.	RRR Construction Co., Inc.
Comprehensive Environ Serv., Inc.	N. Hall Contracting	Selecto-Flash Safety, Inc.
Devo Fire Protection, Inc.	Precision Combustion Consulting	Spoon Electrical Contractors
Executive Abatement Industrial	Program Unlimited Plumb. & Heat., Inc.	Superman Contracting Corp.
Fong & Associates, Inc.	Promatech, Inc.	TCE Systems, Inc.

Figure 7.1. Corporate Membership in the Regional Alliance for Small Contractors of New York and New Jersey (as of Spring 1996)

"majority" firms) have participated in alliance programs during its first 5 years.

The Alliance Forges Relationships in the Community

The alliance long ago extended its reach beyond the traditional players in the world of construction. Its staff and volunteers established working relationships with such training providers as community colleges (e.g., Essex Community College in Newark) and community organizations (with especially close relationships to Newark's New Community Corporation [NCC]). These relationships, for example, connected trainers and trainees from Essex and NCC to job referral and workforce development specialists within the Office of Business and Job Opportunity at the Port Authority's World Trade Center headquarters in Manhattan.

Less conventional interactions followed between the alliance staff and NCC. Cruz Russell, a cocreator of the alliance and currently Acting Director, Office of Corporate Policy and Planning in the Port Authority, provided a support letter on behalf of the Port Authority that contributed to NCC receiving a federal grant to create a subsidiary construction firm. Alliance leaders also subsequently facilitated a meeting between management staff of NCC's new construction entity and hospital construction supervisors and related prime contractors in South Jersey.

Problems and Challenges With Replication and Expansion

Case writer Winter observes that the alliance and its networks did not necessarily achieve success in every area of endeavor. The use of loaned executives to assist firms in appealing minority certification determinations proved to be a disappointing experience for numerous mentors and unsuccessful firms. Efforts to connect minority companies to bank lenders offering attractive terms also fell short of the aforementioned FISC committee's goals, which had been an integral component of the alliance's programs. This performance shortfall may eventually be overcome by increasing bank competition to find emerging minority business borrowers now that the revised federal Community Reinvestment Act requires dramatically increased scrutiny of small business lending in neighborhoods of color. Winter also notes that the Supreme Court decisions in the *Adarand Constructors, Inc. v. Pena (1995)* and *J. A. Croson Company v. City of Richmond* (1989) cases may change the operating approach vis-à-vis inclusionary programs for "disadvantaged" business owners.

In recent years, many of the original contractors, such as Morse Diesel and Bechtel, have endeavored to replicate the alliance approach in other cities (informally in some and more formally, using the name Alliance, in others). In 1995, the alliance also conducted a number of networking functions beyond the Port Authority's target area borders, introducing alliance firms to hospitals in the Philadelphia and Camden area that were planning substantial construction projects. Alliance activities were also initiated in 1995 in Camden and Atlantic City; it is hoped that these will result in the establishment of a permanent program in southern New Jersey.

In 1995, as such Port Authority officials as Cruz Russell began working with the BOC Network, "craft" sharing techniques regarding replication have increasingly been exchanged between BOC leaders and the executive director of the Regional Alliance, Mark Quinn. This interaction between the alliance leadership and the BOC organizers has focused principally on organizational operations issues and techniques to foster successful program replications. Both programs' senior leadership seem to find much common ground to explore, despite the fact that the Regional Alliance activities concentrate on construction and related industries, whereas the BOC constituents principally are smaller scale retailers, wholesalers, and distribution firms. Despite the differing target constituencies, future alliance replications may parallel the BOC approach of providing federal replication support grants, screening the program management capacity of organizations applying to replicate alliance initiatives, providing applicants with a handbook concerning "best practices," access to important human resources, and even office configuration and computer advice.

Alliance director Quinn continues to evaluate how other programs have overcome frustrations when operational approaches are "borrowed" by other jurisdictions. Where such communities fail to emulate key founding characteristics of programs such as the Regional Alliance, they often encounter failures at achieving the results of the original program. Some of these less formal replication efforts have proven unsettling to the reputation of the Regional Alliance because no licensing or quality control agreements were established with communities attempting to initiate some features of the program established by the alliance.

The Regional Alliance as a Model for
Doing "Affirmative Action" Without Quotas

The Clinton administration is currently engaged in restructuring federal emphasis and techniques regarding such inclusionary programs. For instance, the U.S. Small Business Administration's 8A Program for minority vendors to federal agencies has recently been opened up to community development corporations (CDCs). Because the alliance is not structured as a race-based remedial program (it accommodates but does not aggressively recruit nonminority firms), it is likely to become a model for innovative approaches to assisting targeted firms and communities.

COMMUNITY COLLEGES AS INTERMEDIARIES: THE CASE OF LAWSON STATE

For many years, 2-year community colleges have performed the vital function of providing education and training opportunities for those younger people who do not go on to a 4-year college after leaving high school—a pool that still constitutes approximately 54% of the population over age 16. Many of these students go directly into the world of work after completing their programs, whereas others go on to 4-year colleges and universities. The community colleges have also been an educational resource for adults returning to school, retraining for job changes, or acquiring credentials for moving up within their existing companies or agencies.

It is also common for companies, nonprofits, and government agencies in a locale to make contracts with one or another community college in the vicinity to provide specialized ("customized") training programs for their employees. Others enter into "co-op" arrangements, according to which students get to work part-time even before they receive their diplomas or certificates. Such relationships between the 2-year colleges and the automobile companies, for example, have been in place for many years (this is an example of the more general case of labor unions being an important promoter of such collaboration). Often, the businesses will lend or simply give equipment to the schools or offer their own personnel as part-time or temporary teaching staff to simulate within the community colleges the actual workplace environments into which students will eventually move.

These paths into, through, and beyond the community colleges are the ones that educators, especially, think of as the primary mission of their institutions. They are depicted on the right-hand side of Figure 7.2.

Other contractual relationships to specific companies, however, have increased rapidly in recent years, under which the community colleges provide specialized technical assistance to help upgrade the production and management capabilities of local firms. These activities are called many names, such as "business outreach," "industrial modernization," "industrial extension," or "economic development." Former U.S. Department of Labor Secretary Robert Reich emphasizes the importance of this activity to enhancing and maintaining the global "competitiveness" of American industry. James Jacobs, Associate Vice President for Community and Employer Services of Macomb Community College outside of Detroit and one of the most influential innovators in this field, refers to these activities as comprising the "invisible college." He believes that their share of overall community college activity will continue to grow as the more traditional "inside the classroom" education shrinks in scale and in claims on resources.

David Goetsch, President of Vocation and Technical Education at Okaloosa-Walton College in Niceville, Florida, has a more localized perspective. As he expressed it to Dr. Joan Fitzgerald, Assistant

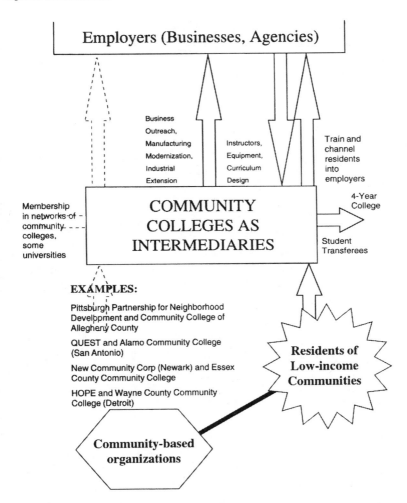

Figure 7.2. Community Colleges as Intermediaries in Workforce Development and Industrial Modernization Programs

Professor of Urban Planning and Policy and Faculty Associate of the Great Cities Institute at the University of Illinois-Chicago (author of our case study on Lawson State Community College), the long-term viability of the community college requires a healthy local economy so that students have at least the choice to find work closer to home when they graduate. These recent relationships between community colleges and particular companies and other employers, aimed mainly at transferring technology and management skills to the firms, are depicted in the link in the middle of Figure 7.2.

Links Between Community Colleges and Community-Based Organizations

Some CDCs and other community-based organizations (CBOs) already work collaboratively with the community colleges in their areas. Although such collaboration is never quite as easy to achieve in reality as it appears in the writing of the "experts," more CBOs should probably be exploring such network possibilities, recommends Dr. Fitzgerald. Community colleges are especially well positioned to help the CBOs expand their job training and placement capacities (Fitzgerald & Jenkins, 1997). Moreover, because companies commonly report that they trust community colleges to a greater extent than they do many other public institutions (an outcome of those long-standing relationships), CBOs that partner with community colleges are potentially buying into networks that can further enhance their own reputations with the private sector.

Earlier, we reported on several evolving CBO-community college-employer networks. Three of these are depicted on the lower left-hand side of Figure 7.2: the long-standing contracts between the Pittsburgh Partnership for Neighborhood Development (or its member CDCs) and various campuses of the Community College of Allegheny County; Project QUEST's engagement of Alamo Community College in San Antonio; and the programs that Essex County Community College operates with and for New Community Corporation in Newark. Another such collaboration that has been much praised (but is not one of our cases) is that between Focus HOPE in Detroit and Wayne County Community College.

"Networks of Networks"

Recently, community colleges have been entering into collaboratives with one another, with the private sector, and with colleges and universities, sometimes across state lines (Rosenfeld, 1995b;

Rosenfeld & Kingslow, 1995). Perhaps the most prominent, but by no means the only example of what Fitzgerald calls "networks of networks," is the Consortium for Manufacturing Competitiveness, which at last count included 15 community colleges and their local partners situated in 14 states, stretching from Texas to Virginia (Rosenfeld, 1994). Figure 7.2 depicts these links.

Lawson State as a Network Member

Lawson State Community College, one of the historically black colleges in the South, is located in Birmingham, Alabama. It provides courses of recognized quality in registered nursing and drafting along with the traditional programs common to schools with this history, such as cosmetology and barbering. Lawson State suffers from a generally poor reputation with companies in the area that is reinforced by its poor funding and strong competition from neighboring colleges and technical institutes, which are (not coincidentally) largely white. For example, the Bevill Center for Advanced Manufacturing, located in nearby Gadsden, is a joint venture of its city, Gadsden State Community College, and the University of Alabama (UA). Bevill and UA-Birmingham have developed far closer and more successful connections to companies; in 1993, Lawson had technical assistance contracts with only 5 firms and in 1994 with only 10, whereas Bevill was working with more than 100 companies on 217 projects.

Like so many other disadvantaged organizations, Lawson State has a history of taking on too many projects and entering too many agreements with others in its quest to bring economic and workforce development to a constituency to whom it feels a powerful commitment and in an effort to obtain additional resources. For example, Lawson participates in a local technology transfer consortium that had been set up by the local public sector to provide technical education to the future employees of the new auto assembly plant that Mercedes-Benz has been constructing outside of Birmingham. Currently, however, Benz plans to conduct its own training in its own facility and is barely aware of Lawson's existence. Partnerships with the chamber of commerce and with several high schools in the city have provided few payoffs for Lawson (beyond goodwill) while only placing further stress on its overworked staff.

Lawson State was invited into the Consortium for Manufacturing Competitiveness (CMC) in 1992 and is the only black institution in the network. By mutual agreement of Lawson's president and of other members of the consortium, there has not been much to show for it. Once again, Lawson's leadership mistakenly thought that the CMC would painlessly bring it additional financial resources, and Lawson was not prepared to dedicate staff time and energy of its own to the collaboration. Its presence at monthly meetings of the consortium gradually dwindled, which only made its own inadequate capacity that much more apparent to the other member community colleges and to the private sector.

From all these experiences, Fitzgerald concludes that, for Lawson State, nonstrategic networking has not relieved its limited organizational capacity, allowing it to expand its range of developmental activities. Indeed, Lawson's approach—enter everything to which it has access and spread the workload among existing staff—has only diminished its effectiveness (and reputation) even further. Lack of staff, inadequate funding and equipment, poor academic reputation, competition from other institutions, and—most of all—lack of programmatic focus have painted this community college, a vital resource to Birmingham's black community, into a corner.

Reinventing Lawson State's Approach
to Network Participation

One of Fitzgerald's recommendations is that Lawson become far more selective in its approach to partnering with others. Closer cooperation with the high-quality and widely admired Bevill Center (another member of the CNC) seems sensible. A second, and more fundamental, suggestion is that the president, staff, and their foundation and state sponsors provide closer mentoring aimed at developing the college's ability to identify a clear niche for its services. For example, with so many local small and medium-sized businesses complaining about the inadequate literacy or numeracy of black youth in the region, Lawson State might well enhance its reputation with the private sector by focusing on basic skills training. As the college's staff acquire competency in this field, they can then move on to other projects.

In a related vein, an "embedded" African American-led institution such as Lawson State might also try implementing a division of labor between community colleges and CBOs. The latter might work especially on soft skills, "code switching" between the norms and rules of the workplace and those of the street, mentoring, and postplacement follow-up, whereas the former emphasize the more formal technical training. Community "ownership" of Lawson matters, to the extent that such a division of labor puts a premium on constant communication among the leaders of both types of organizations.

Finally, Fitzgerald observes that Lawson State does not in fact function as an intermediary between CBOs in Birmingham and the private sector, as we had hypothesized in our initial framing of the fieldwork—but it could. In particular, there is a strong need in the area for additional skilled construction workers, for minority-owned construction companies (of just the kind that NCC and the New York and New Jersey Port Authority have jointly developed in Newark), and for housing and community redevelopment in the wake of the closure of the old U.S. Steel works. Lawson is in fact already loosely connected to the Birmingham Housing Authority and to the principal redevelopment organization, with its roots in the African Methodist Episcopal Church. This cluster of activities would seem to be a promising target for the college to prioritize. The need is great, the community organizations would value the assistance, Lawson's capacity would surely be enhanced by working in the housing sector, and private business leaders would certainly take notice if Lawson succeeded in adding to the city's pool of trained construction workers and capable small construction firms.

Whatever Lawson's leadership chooses to pursue in the years ahead, Fitzgerald concludes that the Lawson State experience to date offers a cautionary tale, a reminder that

> Effective networking does not just happen by joining a consortium. It requires strategic planning to choose the most appropriate networks to create true win-win partnerships. Once the appropriate

networks have been chosen, an institution must be prepared to commit considerable resources in order to reap any benefits. Finally, networking typically takes a sustained commitment before benefits are realized.

SYNTHESIS AND CONCLUSIONS
Toward Better Design, Promotion, and Evaluation of Community-Based Workforce Development Networks

> But your mind usually goes blooey when it's faced with the patterns within more than a dozen or so cases.
>
> —Comment made by a prominent professor at Harvard's Kennedy School of Government on reviewing a new book

The subtext of these studies has been to get us beyond the mantra or buzzword "networking." We are interested in more than the exchange of business cards at conferences or in momentary alliances struck during municipal budget crises or school board elections. What are formal and informal networks? Can theories of contractual and social relationships help us to understand them? Who initiates them and in what circumstances? How do they work? What do they do? How extensively are they being used?

At least with respect to the subject of workforce development, thanks to the sustained and imaginative research commitments from major funders during the past 6 years, we have made major breakthroughs in getting answers to some of the large, long-standing puzzles in this field. For example, the fieldwork by Clark and Dawson, Mt. Auburn Associates, and our own research on "building bridges" strongly supports the conclusions reached by Tom Dewar and David Scheie (1995) at Rainbow Research, Inc., on the importance of designing workforce development efforts with "strong ties to real employers, hands-on training, life skills along with technical skills, instructors with real industry experience, and self-paced learning" (p. 44). We now know that "ongoing relationships with industry insiders are critically important," that it is imperative (especially when working with adults) to "combine technical or vocational skills with broader life skills and support," and that follow-up mentoring and counseling are both crucial and constitute the element that mainstream programs can least afford (or are least inclined) to pursue (pp. 84-85).

Probably our own single strongest finding is the central importance to community-based organizations (CBOs) of (in Edwin Melendez' words) becoming an ongoing, trusted part of the recruiting and training networks of a region's employers. In such best-practice program designs, companies actively participate in identifying and selecting occupations and skills for which training will be provided and in designing the training programs. Executives regularly visit CBO training sites. Instructors, often recruited from the very companies that are doing the hiring, develop (if they do not already have) a personal stake in seeing trainees succeed. On this score, the Center for Employment Training (CET) is the acknowledged model for its commitment to addressing what labor economists call the "demand side" (employers) as well as the "supply side" (workers and trainees) of the market. This can also be seen in Coastal Enterprises' system of writing formal hiring contracts with companies, enabling the community development corporation (CDC) to offer training with a high probability of a real job placement at the end. This is also central to the work of San Antonio's Project QUEST.

Similarly, there is convergence among researchers that narrow specialization on one or another element of the workforce development process is almost always inadequate. Rather, there is a need to package outreach, recruiting, training, placement, follow-up counseling, child care, and transportation. "Packaging" is different from—albeit not inconsistent with—agencies or governments attaining efficiencies or raising trainees' "comfort levels" through so-called "one-stop shopping." Although expensive, Project QUEST's comprehensive packaging of services, along with the provision of training stipends, speaks to a need that the greater bureaucratic efficiencies offered by one-stop shopping by themselves cannot offer. Indeed, it is precisely this need to package elements into a "system" that so compels groups to network with one another and with mainstream institutions.

So much remains to be learned, however. For example, future research should be directed to answer the following:

◆ Why and how do CDCs and other CBOs actually start up (or enter existing) networks?

◆ How stable, durable, and long-lasting are these network relationships?

◆ How do network managers deal with mutual responsibility, accountability, and governance?

◆ Compared with the theory, and with initial expectations of the actors, what does networking actually "buy" the organization that enters into it?

◆ What about area "spillovers?" What does networking do for the neighborhood, community, and region beyond the immediate network members, such as seeding other agencies with talented, experienced staff?

◆ How is "exit" from a network (the termination of a networked relationship) handled? When is it deliberate or unintended?

◆ What is the relative importance of politics and culture in network formation and governance? In particular, do the CBO and its partners display a particular political strategy as to which segments of the residential community and the region's businesses they want to work with?

◆ How (and how well) are the networks staffed? What is the mix of full-time paid staff, part-timers, and volunteers? Do at least some of the member CBOs (or a governing committee or "central office") provide personnel whose time is substantially dedicated to network management?

◆ What difference does it make? Does the network have dedicated services, equipment, and real estate?

◆ In the interest of assisting CBOs and promoting community development, what can or should the public sector, the private sector, and the philanthropic foundations do to help these networks operate?

Some of this unfinished business is conceptual, and some of it is evaluative. Here, we examine more closely several of the still unanswered (or inadequately addressed) conceptual issues. Problems with and challenges to evaluation will be discussed later.

THE LEADERSHIP QUESTION

Some CBOs have survived for a very long time under the same charismatic (often white) leaders. Among our cases, this has been true of CET (under Russ Tershy), New Community Corporation (NCC; under Father William Linder), and Bethel New Life, Inc. (under Mary Nelson). Other groups, such as Communities Organized for Public Service (COPS), are an integral part of organizing networks (in COPS' case, the Industrial Areas Foundation) that make the training and development of new leadership a centerpiece of their work.

When one spends time with these groups, one is continually struck by the skill and depth of their staffs. The prominence of the top leaders should not mislead us into overlooking the richness of the cadres of community development professionals of color whom these and other organizations have nourished. When NCC's vice president for human resources, Florence Williams, describes the country's largest CDC as "seeding" the rest of the city and the region during a period of 20 years with trained personnel, from program operators to bus drivers, she is not exaggerating.[1] Moreover, NCC and CET (to name only two of the groups we have studied) are now devoting substantial resources to staff and leadership development.

The subjects of leadership training and succession take us back to a much larger question that was raised by veteran organizer Marshall Ganz. When CBOs are successful (if only in surviving the buffeting of hostile environments), how much of this success is attributable to agency and how much to institutions? As was previously shown in summarizing the case studies on the two Hispanic organizations CET and QUEST, Ganz sees the interplay between these dimensions as critical to making sense out of how they work. This is an area of study in which the community development and employment training and workforce development movements should be able to engage the interests of organization theorists, to their mutual benefit.

THE PARADOX OF NETWORKING
AND ORGANIZATIONAL CAPACITY

In this book, we have on several occasions encountered a paradox about networking as a way around inadequate organizational capacity. On the one hand, entering networks is a way of accessing capabilities or capacities that any particular focal organization does not have or cannot afford. On the other hand, even a small organization needs certain capabilities or capacities to be able to intelligently scan the environment and decide which workforce development networks are worth joining. This need to make choices and govern complex relationships grows geometrically with networking.

How are CBOs or intermediaries approaching this dilemma? Not always wisely, according to our case studies. Some (such as Lawson State College) enter too many networks, imagining that their own internal organizational capacity constraints will be relieved by the network partners. Others, such as Bethel New Life, not only are extremely "thin" when it comes to devoting internal staff to the management of network relationships with outsiders but also seem disinclined to partner with precisely those nearby community groups that might best complement their own staff resources. Thus, there is much research to be done on what kinds of internal organizational reforms, especially regarding management of staff, might help the CBOs and intermediaries to more effectively engage in external network activities.

In truth, the problem of inadequate attention to continually honing and expanding organizational capacity is a chronic malady of community development organizations in general. In that sense, the workforce development networks we have been studying are no exception. Thus, in a recent study of Local Initiatives Support Corporation's 17 Operating Support Programs across the country (programs aimed precisely at enhancing CDC capabilities in terms of housing development and management), Nancy Brune, Yale University law student, reports that

"one OSP director noted . . . that organizational and board development 'was only given lip service.' Another said: 'Organizational development has been overlooked almost entirely. If it happens, it's only a
by-product' " (Brune & Bylenok, 1996, p. 12). Failure of a CDC to
actually build housing, however, could be cause for suspension from the
quite lucrative program.

COMMUNITY-LABOR COALITIONS
(AND WORKING WITH LABOR UNIONS IN GENERAL)

Relationships between community development groups, community
organizers, the civil rights movement and labor unions have a long and
uneven history. It is incontestable that workers of color earn the highest
wages and enjoy the most extensive benefits in the most highly unionized
industries and occupations (including the federal civil service). Particular
unions, notably the United Auto Workers during the 1950s and 1960s,
were genuinely committed to civil rights and promoting the fortunes of
inner-city CDCs. Many veteran community organizers, however, can
remember facing craft and other unionized white male workers across
barricades, weapons in hand, contesting rather than sharing (and,
through their combined strength, adding to) the scarce supply of construction and related jobs. In recent years, as the economy has been
restructured away from basic manufacturing and toward services, with
the advent of interracial community-labor coalitions confronting factory
closures, and with the growing numbers of African American, Latino,
and Asian members within the unions, long-standing walls between
community development activists and the struggling-to-reinvent-itself
labor movement may at last be coming down.

Among the 10 cases, we found few examples of ongoing partnerships
between CBOs or intermediaries and unions. There certainly have been
opportunities foregone. One example is from the West Side of Chicago,
adjacent to the "turf" of Bethel New Life. There, a rich mix of institutional actors is currently working (more or less collaboratively) to
revitalize the metalworking district in and around the now mostly
African American neighborhood of Austin. With its population of

approximately 100,000, even after many years of deindustrialization, this area (like others in and around the city) retains a mass of mostly small and medium-sized metalworking shops owned almost entirely by white ethnic men, mostly of Eastern European origin. Between the aging and retirement of the white working class of machinists who historically worked in these firms and the continuing suburbanization of the area's white population, these owners increasingly find themselves facing a chronic shortage of both skilled and apprentice workers for a set of occupations that pay above-average wages and are in demand in cities across the country.

Only approximately 9% of the more than 90,000 employees in these 4,000 businesses are residents of Austin and its environs. Moreover, these small shop owners are confronted with the common problem of "successorship," with few of their children or other relatives interested in eventually taking over these businesses, certainly not in their current location. Finally, to remain competitive as suppliers to the big Chicago area customer firms in auto, farm, and communications equipment, let alone to become exporters to other regions, these small shops need to undergo further technological upgrading and restructuring of their management practices.[2]

For almost a decade, a local activist working out of the historically famous Chicago settlement house movement, Ric Gudell, has been slowly familiarizing the owners of the small shops in the Austin metalworking sector with the potential for hiring African American employees from the neighborhood. Recently, under the sponsorship of the South Shore Bank, an Austin Labor Force Intermediary and Austin Enterprise Center have been created to facilitate the simultaneous attention to upgrading the technical and managerial capacities of the companies in the area. The bank is probably the premier model in the country of how an urban development bank might operate in the joint interests of its investors and its community.[3] The project is also associated with two citywide coalitions of CBOs, the Chicago Jobs Council and the Chicago Association of Neighborhood Development Organizations. Research and community organizing are also contributed by the Midwest Center for Labor Research, an independent activist "think tank" with strong ties to the labor movement and an impressive track record in working with communities of color.

With financial and technical support from city government, the banks, private foundations, and universities, Chicago's Austin Initiative, although still very new, clearly nests within and draws on a rich web of organizations. Even the federal government is involved, with the National Institute of Standards and Technology (NIST) in the U.S. Department of Commerce having created the Chicago Manufacturing Center—the newest node within NIST's national network of 44 Manufacturing Extension Partnerships in 32 states (Kelley & Arora, 1996; Sabel, 1996). The center has already begun to insert itself into this matrix of enabling institutions on the West Side of Chicago.

Although it is situated in the center of the action, Bethel New Life has chosen not to be a major player within this set of intersecting networks. As the case on Bethel indicates, the CDC has at most an informal relationship to the other actors (except for its preexisting partnership with Argonne National Laboratory) and no connection to Chicago's unions at all. The former labor organizers from the Midwest Center have been especially keen to arrange such a partnership, but for one reason or another, Bethel has kept its distance from what can only be called a promising opportunity.

Community-labor coalitions admittedly have had a checkered history in American cities. One prominent failed effort to build this particular bridge took place in Boston during the 1980s when activists attempted to link a well-known Roxbury CBO, the Dudley Street Neighborhood Initiative (DSNI), to local union activists. During the summer of 1986, DSNI entered into discussions with Local 26 of the Hotel and Restaurant Employees International Union about possible collaborations. Local 26 had a history of progressive organizing, and its membership increasingly consisted of Latino workers, many of them recent immigrants. Local 26 had identified affordable housing as a priority area, with many of its members already living in the Dudley Street area (Medoff & Sklar, 1994).

There was certainly a basis for collaboration. Both organizations were multiracial and ethnic, progressive, and organizer-driven. Whereas the CBO could add to the union's legitimacy in the neighborhood, Local 26 brought to the table its assistance in connecting DSNI's constituents to jobs in downtown hotels and potential access to the union's substantial reserve of pension funds for financing affordable housing. As a good faith "down payment" on the collaboration, DSNI members joined in a series of local actions against the Back Bay Hilton, then a nonunion hotel, in support of Local 26's organizing drive.

This promising alliance foundered, however, when Local 26's leadership first joined forces with a prominent Latino CDC, Nuestra Comunidad, with a plan to tap into the pension fund to finance new construction in the area on city-owned land. Neustra was criticized by residents for its design process and for the slow pace with which it was able to move through the bureaucratic permitting process. Interethnic politics added to the problem because a number of African American developers chose to depict any Nuestra-DSNI-Local 26 alliance as an "Hispanic conspiracy."

A more successful alliance has evolved in Baltimore between the largely African American Baltimoreans United in Leadership Development (BUILD), which is the local affiliate of the same Industrial Areas Foundation organizing network of which Texas' COPS is a member, and the American Federation of State, County, and Municipal Employees (AFSCME). Municipal and private business leaders had promised the black churches, whose members constitute the base of BUILD, that the latest wave of downtown construction of hotels and office towers would create many jobs for those parishioners. Too many of the jobs, however, ended up paying poverty-level wages and providing few or no benefits. Meanwhile, the experienced public employees in the city found themselves being shut out of the employment boom because of the city's practice of contracting out much of the work to private low-wage firms that would hire—whom else?—the black members of BUILD.

In 1993 and 1994, an alliance between AFSCME and BUILD eventually pressured the (black) mayor, the city council, and the downtown developers into legislating what unions call a "prevailing wage" ordinance, now popularly called the "living wage," setting a floor under the hourly pay of all contracted workers. The union estimates that the signing of this ordinance in December 1994 meant an immediate 40% increase in the earnings of employees in the contract shops while somewhat reducing the incentive to the city and the developers to try to escape union labor by contracting out (Keler & Lange, 1995; Peschek, 1997). Living wage campaigns based on the Baltimore model are currently being organized in cities throughout the country, mostly under the leadership of the Association of Community Organizations for Reform Now, or ACORN.

BENCHMARKING

Taking stock of best practices by similar organizations in a field—the problems they confront, their responses to them, how well they manage, and what peers think of them—has become a standard part of the repertoire of management consultants, selling high-priced advice to the world's largest corporations. The CDCs, other CBOs, and intermediaries we have been studying, in some cases for as long as 5 years, vary enormously in the extent to and seriousness with which they "benchmark" their peers, competitors, and collaborators. CET, QUEST, and NCC engage in extensive benchmarking. Bethel New Life and Lawson State College do not seem to do much of this.

For the purposes of self-evaluation, in planning the replication of initial efforts at other sites ("cloning," "going to scale," and so on) or just to remain on top of the field, comparative benchmarking is crucial. Because collaboration with other organizations through both informal and formal networks is both more complex and rather more "experimental" than more conventional approaches to meeting such goals as workforce development, conscientious benchmarking is especially important. Assisting CBOs to conduct benchmarking exercises, if only for their own internal use, is one activity on which outside advisers, graduate students from local colleges and universities, and former staffers can often offer useful "technical" assistance.

Coastal Enterprises, Inc. (CEI) and the Business Outreach Center (BOC) network utilize benchmarking techniques that extend beyond their internal performance review to assess the impacts of network contacts on the businesses they have targeted for capital or technical assistance support or both. CEI uses surveys to obtain feedback from entrepreneurs regarding the employment experiences of former welfare recipients who have been placed in private-sector jobs in companies receiving loans from CEI. The BOC network also surveys its business clients on a regular basis. BOC focuses its inquiry on assessing the quality of resources provided by public, nonprofit, and private firms to which the entrepreneur has been linked by the BOC. The BOCs then "vote with their feet" by avoiding future referrals to organizations and institutions deemed inadequately responsive to previously referred client business. Eventually, funding or regulatory sources become aware of the fact that small businesses from specific poor neighborhoods are being referred by the BOC network to another part of the city because business technical assistance or smaller loans were deemed to be inadequately provided by a more localized small business development center or bank.

Benchmarking is actively encouraged by one of the CDCs' most important sources of operating grants (a kind of "venture capital") from the federal government. These are the grants made each year by the Office of Community Services (OCS) of the U.S. Department of Health and Human Services. In connection with these operations grants and funding for its Job Opportunities for Low Income Individuals program, OCS has also operated an innovative Demonstration Partnership Program since 1987. Most of these demonstrations are sponsored by CBOs seeking to assist local welfare recipients to make the transition to private-sector jobs. The agency makes available to its nonprofit grantees several helpful benchmarking and "lessons learned" manuals.[4]

In these manuals, two particular demonstration projects are highlighted. One award-winning program run by the community action agency in Pierce County, Washington, with OCS funding created an extensive postplacement support activity, the Steps to Career Success Program. It offers assistance to newly employed former welfare recipients in dealing with crises such as urgent car repairs, work clothing, child care services, rent money, and health emergencies. Other intervention services include support counseling, referrals, networking, life career planning, and "needs-solving" skills. The STEPS program extends its benchmarking by evaluating a comparison group without comparable interventions located in a nearby city. OCS's preliminary findings suggest that STEPS clients reverse the usual recidivism experience of the state of Washington regarding welfare to work (32% after 12 months and 74% after 26 months). STEPS program coordinators report that STEPS clients "move ahead on their jobs, receive more promotions, have a lower rate of absenteeism, are more likely to seek counseling for issues which could be potential barriers to employment and have a higher rate of enrollment in further education" (from "Steps to Career Success," news release upon being named Outstanding Rural Development Project by the National Association of Community Action Agencies and the National Association of Development Organizations Research Foundation, April 3, 1996).

Five years ago, when we first began our site visits with CDCs and CBOs interested in connecting low-income residents to permanent jobs, the issue of preplacement "world-of-work" orientation was raised as another critical element of local workforce development. An OCS Demonstration Partnership awardee in Chicago operates the Two Generation Head Start Self-Sufficiency Program also known as "Step-Up." In conducting an evaluation of Step-Up, David Beer, from the University of Illinois at Chicago, noted that the program's purpose was to design and implement a full range of job preparation supports and activities for Head Start parents. The Welfare-to-Work project was conducted in collaboration with the City of Chicago's Department of Human Services, the Erikson Institute, and the Holy Family-Cabrini Head Start (Project Match) program, with the latter serving as the demonstration site. Clients were provided services regarding assessment, initial placement, retention, re-placement and other job-related supports—the most important being regular biweekly or monthly counseling, comparable to that provided in San Antonio by QUEST.

Beer reports that Project Match/Step-Up recognized that the path from welfare to work entailed complicated steps rather than merely crossing a "boundary." Therefore, the program offered stepping stones for less job-ready parents based on volunteer activities in local organizations and stipend support for other Head Start assignments. The evaluator notes the importance of the Project Match pre- and postplacement tracking systems and provides data demonstrating the positive impact the Step-Up program was having on clients. Ironically, due to congressional budget cuts targeting discretionary spending support for continuation of the demonstration partnership, this program—which OCS uses for benchmarking welfare-to-work planning, another goal of Congress—is scheduled to expire in the near future.

QUALITATIVE AND PROCESS ISSUES
IN PROGRAM EVALUATION

Guidelines for designing formal evaluations generally fall into four categories: (a) evaluating organizational capabilities and the growth of capacity, (b) identifying and measuring training "inputs" directly attributable to the project, (c) identifying and developing indicators for measuring project outcomes, and (d) conditioning estimated outcomes on the institutional, social, and local context. The technical evaluation field has come a long way during the past quarter century and now contains a number of large specialist firms and a cadre of individual scholars who regularly produce high-quality work, especially with regard to elements b and c. The next step in this evolution will be greater sensitivity to and development of enhanced ability to measure elements a and d.

An example of the first element that tends not to be given much attention in mainstream evaluations, but is very important, is measurement of the extent to which the organization invests in staff development and is able to promote qualified staff—giving them enhanced responsibilities and commensurately higher salaries—either within a particular CBO or somewhere in the larger network. Moreover, even when qualified staff "exit" the network, whatever "demerits" are assigned to the original CBO for "failing" to retain its staff should be offset by whatever contributions that "lost" employee is making to the community development effort somewhere else—as in Chicago and Newark, where former Bethel and NCC staffers can now be found working in other CBOs in other low-income neighborhoods. The loss to a particular organization of a qualified staff person who became more skilled as a result of working in that organization may still be adding to the capacity of the larger community—to the pool of cadres of skilled community organizers, activists, and program operators. If so, the original CBO should get some credit for that staff training and capacity enhancement. It will be expensive for evaluators to track mobile staffers, but we think this is an important future direction for research.

With regard to measuring organizational growth, there has been hardly any progress, even though funders and activists fully recognize that this is in some sense what social programs are mainly about. In this era, everyone wants to be thought of as "acting strategically," and CBOs and their networks should be assessed for how well they anticipate, plan, and deliberately choose certain paths. At the same time, it is also a great skill and not necessarily a put-down to be known as "creative" in the sense of leadership seizing unanticipated chances. Measuring such capabilities will be horrendously difficult (because it is subjective). All successful organizations, however, grow, survive, and recover from crisis in part because their leadership has skill at just this ability to take advantage of ("run with") unexpected openings.

Existing evaluation research hardly addresses context—element d. It is impossible, however, to fully understand the sources of the legitimacy of groups such as CET on the West Coast or QUEST in San Antonio, why the Hispanic population trusts them, and why employers pay them so much attention and respect without explicating their histories. CET's experience in working with farmworker organizers is, we are convinced, a key to understanding its successes. Similarly, QUEST is immediately identified (by ordinary people on the street, by bankers, and by city and state officials) as a project of COPS, and therefore of Texas Interfaith and of Industrial Areas Foundation (IAF). It is deeply, and fairly transparently, embedded in an active community organizing movement. To the extent that COPS' replication of QUEST in other areas grounds those extensions within the IAF network (in ways that the far more ambitious and CET replication has not been able to do thus far), QUEST's chances for future success are, we predict, enhanced.

Even elements that are already a standard part of conventional program evaluations need to be rethought. An example of the need to rethink element c, the measurement of program outcomes, arose in the case study of QUEST. QUEST is, by design, an expensive program. Given its chosen emphasis on credentialing and retraining somewhat older, often displaced workers, it must provide a comprehensive package of benefits. This means, however, that the measured benefits from and payoffs to the investments in training (measured, for example, by preintake vs. postplacement earnings growth vis-à-vis some control or comparison group) must be substantial to justify the expense. What will make it even more difficult for QUEST to show high benefit-cost ratios is the way in which trainees are selected in the first place.

As we have seen, residents often learn about QUEST through their engagement with the Hispanic churches or with COPS' and the Metro Alliance's various chapters and groups. Applicants are then scrutinized by fellow citizens acting on behalf of COPS-Metro who not only can influence and guide the applicants but also have the power to reject them—to screen them out on the grounds that the applicant seems not to be ready to undertake the commitment of time and energy to complete a rigorous program of as much as 2 years' duration. To increase the likelihood of success even further, applicants who make it through this first screen then get tested and evaluated on literacy and numeracy skills. The net result is that QUEST clearly does (to use a well-known jargon term in the field) "cream" the applicant pool. Evaluators will normally take that into account and, using one method or another, discount the measured benefits and payoff on the reasonable grounds that these successful graduates would have done better than average anyway, even in the absence of the QUEST "treatment." This makes it that much harder for a program such as QUEST to show a high rate of return to the investment.

QUEST's creators, however, argue that the screening process, especially its first stage, empowers ordinary citizens in the community precisely by giving them such a central role in the process. They are being given power, responsibility, and—because neighbors are judging neighbors—there is accountability as well. Whether or not a conventional benefit-cost analysis treats such a program design as being "not successful" from the perspective of community building, the design may be very successful indeed. In short, as important as they are as goals, improving postplacement employability and earnings are insufficient ways of understanding the objectives of QUEST. When these are redefined to include expanding the confidence and skill of neighbors at screening prospective trainees, evaluators could very well conclude that, on balance, the program is a "success." It depends on how the program objectives are defined and how well the evaluators incorporate them into the analysis. In fact, in his recent (first ever) formal evaluation of

QUEST, Massachusetts Institute of Technology's Paul Osterman has been quite mindful of these larger issues.

PROMOTING LEADERSHIP AND STAFF
COMMUNICATION AMONG ORGANIZATIONS

The Clark and Dawson (1995) monograph reports that all the programs its authors studied "developed in complete isolation from one another, in . . . response to their local circumstances" (p. 1). This tendency for CBOs to continually be "reinventing the wheel" is an old and frustrating story. Foundations, supportive government agencies, and other institutions attempt to reduce duplication of effort and endless starting over by commissioning reports (such as the one you are now reading), sponsoring conferences, and creating coordinating "intermediaries" at both the local and national levels.

We suggest that a next step in such efforts might be to promote and fund visits and exchanges of CBO staff among one another's organizations and projects. Scholarships and other arrangements might finance the salaries of regular staff of one organization to work for a period of time—perhaps as much as 1 year—inside a sister CBO at another location, with tasks organized around the management and evaluation of networking activities. These personnel exchanges speak to the need for greater interorganizational communication by assisting staff to "get inside the skin" of another group. They are not meant to replace but rather to supplement the support of cadres from any particular CBO that are dispatched by the CBO itself to other sites—for example, to manage replication efforts.

One such effort at institutionalizing cross-CBO "visits" is currently in the exploration and prefunding process at NCC. NCC has crafted the concept of creating a 1-week "campus" for colleagues from other CDCs throughout the country. During the course of the week, the attendees would be exposed to systems and operating challenges that confront the managers of NCC's diverse economic development and human service programs. NCC staff who actually run the Pathmark Supermarket, the nursing homes, the printing business, the home health care initiatives, the day care centers, the credit unions, and the CET training programs would serve as trainers, along with several consultants, in rigorous programs sharing important craft and managerial experiences. Unlike the typical 1-week training programs offered at universities for "mid-career" professionals, the funding approach for this peer-to-peer resource program would provide for 10 to 20 hours of postvisit access to venture managers and training instructors via telephone, e-mail, and homepage information dissemination. NCC is also exploring the compilation of such important information via digitized videotapes, accessible through project-related Internet homepages.

In fact, during the past several years, OCS has initiated and funded a small program that provides peer mentors from various experienced CDCs who deliver technical assistance support to emerging CDCs and other colleagues without charge. Similarly, the U.S. Justice Department and Department of Housing and Urban Development are currently collaborating with OCS in a weed and seed initiative that delivers training and technical assistance via consultants whose work is being coordinated by the National Congress for Community Economic Development (NCCED).

NCCED has become the major Washington, D.C., lobbying organization—the "trade association"—for the CDC movement. In that capacity, it performs a number of invaluable functions. Its research and documentation activities, however, have always, of necessity, been subordinated to its policy and networking roles. We think that funders might usefully promote more specialized and centralized research and practical documentation of workforce development, housing, enterprise creation, and other CDC and CBO activities—including a greater awareness of interorganizational collaborations—by reinventing a national center for just this purpose.

We say "reinvent" because, in fact, just such an entity once existed in the early days of the movement. From the late 1960s through the early 1980s, a central repository of CDC project information and documentation was maintained in the library at the Cambridge, Massachusetts-based Center for Community Economic Development (CCED). Few of the original Title VII CDCs would have considered entering into a new programmatic area without first checking with CCED technical assistance director, Susan Horn-Moo and the CCED librarian, Florence

Mercer Roisman. CCED's mission was undermined in the 1970s, however, and the organization was shut down after Ronald Reagan became president in 1981.

Now, as the movement has grown from the original 40 CDCs to its current scale, numbering in the several thousands throughout the country, new programmatic developers in individual organizations are often forced to learn about previous comparable efforts by attending a conference or a workshop at which three or four panelists are allowed only 10 to 15 minutes to describe the effort of setting up a training program, creating a loan fund, or building a shopping strip. *Building Bridges* (Harrison, Weiss, & Gant, 1995) coauthor Jon Gant suggests that, with the computer's ability to digitize project photographs, videotapes, and working documents, a central repository of detailed programmatic operational documents can be made readily available to any interested CBO within minutes of accessing the Internet. A combination of data availability and on-line responsiveness to project operational questions might be worthy of sustained support from national funders.

ENCOURAGING AND FACILITATING NETWORKING ACTIVITIES AMONG CBOs

There are many things that funders, individual governments, and such associations as the U.S. Conference of Mayors, the National League of Cities, and the National Governors' Association can do to nurture and further promote existing programs. For example, the Regional Alliance for Small Contractors connects prospective small firm suppliers and partners (mostly from the city or region) with big city, suburban, and international firms seeking local or neighborhood partners. The alliance could do even more with the active support of, and be a valuable resource to, mainstream public institutions. Funders can also invest in the capacity and promote the visibility of civic and trade associations that might serve as bridge builders ("facilitators"), provided they are inclusive of businesses and social service organizations from different neighborhoods, races, and ethnicities.

Foundations that currently have Program Related Investment (PRI)-type loan funds could come together and teach what they do and how they do it to consortia of banks and local governments that might be encouraged to create regional "capital funds." The existing individual, within-foundation PRIs, and the proposed regional capital funds could offer incentives (extra points in the scoring of applications and more attractive borrowing terms) to CBOs and businesses that explicitly build networking into their proposals.

Finally, with respect to further promoting workforce development, funders, governments, and trade associations might be encouraged to deliberately create (or favor) training consortia that facilitate locating, recruiting, training, and postplacement follow-up. These consortia could also be engaged in providing development of the training staffs of individual CBOs.

PROMOTING GREATER PUBLIC AWARENESS OF SERIOUS, STRUCTURED NETWORKING ACTIVITIES

Our case studies and our survey of other related efforts throughout the country reveal an abundance of promising activity with regard to interorganizational network collaborations not only regarding workforce development but also in the areas of housing, health care, child services, and transit. Even as they continue to evaluate these efforts, and program operators continue to modify them, it seems long past due for these activities to be made more visible to residents, leaders, and the general public.

There is much that funders can do, under the rubric of "communications policy," to get the word out and encourage more broad-based support for these activities. For example, as with the Urban Engineers program developed (and, it is hoped, continued) by Bethel New Life and Argonne National Laboratory, and as educator Larry Rosenstock has been doing for years at the Cambridge Rindge and Latin School, representatives from these projects can be brought into the public schools, where kids, teachers, and parents can see what is developing. Youth advisers can be mobilized to help ensure that what gets displayed ("performed") at these events really grabs and holds the students' attention.

Funders can help community-based but regionally engaged networks to host regular areawide forums with and among people from community organizations, business, government, unions, the media, and the general public.

Using as a model the national Ford Foundation-Kennedy School of Government (Harvard University) "Innovations" awards for creativity and excellence in the public sector or the Baldridge Awards for technical excellence in the private sector, funders can create award contests to celebrate and publicize achievements by local interorganizational networks and alliances. Finally, national and even occasional international "road shows" might be sponsored to display networking and partnering projects to decision makers in other regions and countries, with the goal being to attract still more prospective partners to an ever-widening circle of collaborators.

NOTES

1. It would be tremendously useful for funders to support research that tracks this seeding process, tracing the careers of "graduates" of community-based programs. Program and project evaluators should be considering such regionwide "human capital" as what economists would call a "joint product" of the activity being evaluated.

2. For a history of the Chicago metalworking districts dating back to World War I, see McCormick (1996).

3. On the history of ShoreBank and efforts to replicate it in other parts of the United States, see Taub (1988).

4. See Rogard-Tabori (1995) and Analytical Database Systems (1995).

REFERENCES

Ackerlof, G. A. (1970, August). The market for "lemons": Quality, uncertainty and market mechanism. *Quarterly Journal of Economics, 488-500.*

Adarand Constructors, Inc. v. Pena, 115 S. Ct. 2097 (1995).

Analytical Database Systems. (1995). *Self-sufficiency project implementation manual.* Washington, DC: Author.

Anderson, E. (1990). Racial tension and job training. In K. Erikson & S. Peter Vallas (Eds.), *The nature of work: Sociological perspectives.* New Haven, CT: Yale University Press.

Appelbaum, E., & Batt, R. (1994). *The new American workplace.* Ithaca, NY: ILR Press.

Applebome, P. (1995, February 20). Employers wary of school system. *New York Times,* p. A1.

Argonne National Laboratory. (1995, June). *Restoring our urban communities: A model for an empowered America: A unique partnership between Bethel New Life and Argonne National Laboratory* [Informational booklet]. Naperville, IL: Author.

Bailey, T. R., & Bernhardt, A. D. (1996, October 2). *In search of the high road in a low-wage industry* (Working Paper No. 2). New York: Institute on Education and the Economy, Teachers College, Columbia University.

Barley, S., Freeman, J., & Hybels, R. C. (1992). Strategic alliances in commercial biotechnology. In N. Nohria & R. Eccles (Eds.), *Networks and organizations: Structure, form, and action.* Boston: Harvard Business School Press.

Bartik, T. (1995, June). *Using performance indicators to improve the effectiveness of welfare-to-work programs* (Staff Working Paper No. 95-36). Kalamazoo, MI: Upjohn Institute for Employment Research.

Bartik, T., & Bingham, R. (1995, January). *Can economic development programs be evaluated?* (Staff Working Paper No. 95-290). Kalamazoo, MI: Upjohn Institute for Employment Research.

Bell, S. H., Orr, L. L., Blomquist, J. D., & Cain, G. G. (1995). *Program applicants as a comparison group in evaluating training programs.* Kalamazoo, MI: Upjohn Institute for Employment Research.

Berger, S., & Piore, M. J. (1980). *Dualism and discontinuity in industrial societies.* New York: Cambridge University Press.

Bergmann, B. (1996). *In defense of affirmative action.* New York: Basic Books.

Berndt, H. E. (1977). *New rulers in the ghetto: The CDC and urban poverty.* Westport, CT: Greenwood.

Bishop, J. (1994a). *The incidence of and payoff to employer training* (Working Paper No. 94-17). Ithaca, NY: Cornell University, School of Industrial and Labor Relations, Center for Advanced Human Resource Studies.

Bishop, J. (1994b). The impact of previous training on productivity and wages. In L. M. Lynch (Ed.), *Training and the private sector: The international comparisons.* Chicago: University of Chicago Press.

Bluestone, B. (1995, Winter). The inequality express. *The American Prospect,* 81-93.

Bluestone, B., & Rose, S. (1997, March/April). Overworked and underemployed. *The American Prospect,* 58-69.

BOC Network Service Impact Survey. (1995, Spring). *BOC Sights, 1*(1), 1-4.

Bosworth, B. (1996). *Using regional economic analysis in urban jobs strategies.* Cambridge, MA: Regional Technology Strategies.

Briggs, X. S., & Mueller, E. (with M. Sullivan). (1997). *From neighborhood to community: Evidence on the social effects of community development.* New York: The New School for Social Research, Community Development Research Center.

Brophy, P. C. (1993). Emerging approaches to community development. In H. G. Cisneros (Ed.), *Interwoven destinies: Cities and the nation.* New York: Norton.

Brune, N. E. L., & Bylenok, D. J. (1996, April). *Balancing support with accountability in operating support programs: A policy analysis exercise for the local initiatives support corporation.* Cambridge, MA: Harvard University, John F. Kennedy School of Government.

Bureau of National Affairs, Inc. (1994, September 26). *Daily Labor Reporter,* p. B-19.

Burt, R. S. (1992). *Structural holes: The social structure of competition.* Cambridge, MA: Harvard University Press.

Burtless, G. (1996, January 1). *Trends in the level and distribution of U.S. living standards, 1973-1993.* Washington, DC: The Brookings Institution.

Cappelli, P. (1995, December). Rethinking employment. *British Journal of Industrial Relations, 33*(4), 563-602.

Cappelli, P., Bassi, L., Katz, H., Knoke, D., Osterman, P., & Useem, M. (1997). *Change at work.* New York: Oxford University Press.

Cave, G., Bos, H., Doolittle, F., & Toussaint, C. (1993). *Job start: Final report on a program for school dropouts.* New York: Manpower Demonstration Research Corporation.

Clark, P., & Dawson, S. L. (with Keys, A. J., Molina, F., & Surpin, R.). (1995, November). *Jobs and the urban poor: Privately initiated sectoral strategies.* Washington, DC: Aspen Institute.

Connell, J. P., Kubisch, A. C., Schorr, L. B., & Weiss, C. H. (Eds.). (1995). *New approaches to evaluating community initiatives: Concepts, methods, and contexts.* Washington, DC: Aspen Institute.

County of Santa Clara (California). (1974, August). *Evaluation of the OIC/welfare contract.* Santa Clara, CA: Author.

Deitrick, S., & Harrison, B. (1994). *The Pittsburgh transition: Planning for economic development in a world of change.* Pittsburgh, PA: Urban Redevelopment Authority.

Development Associates, Inc. (1971). *An evaluation of EDA training related projects: Findings-analysis-conclusions-recommendations.* Washington, DC: U.S. Department of Commerce, Economic Development Administration.

Dewar, T., & Scheie, D. (1995). *Promoting job opportunities.* Baltimore, MD: Annie E. Casey Foundation.

Dickens, W. T. (1996, November 14). *Rebuilding urban labor markets: What community development can accomplish.* Washington, DC: Brookings Institution.

Doeringer, P. B., & Piore, M. J. (1971). *Internal labor markets and manpower analysis.* Lexington, MA: D. C. Heath. (Reprinted with a new introduction, 1985, Armonk, New York: M. E. Sharpe)

Economic trends. (1997, January 27). *Business Week,* p. 20.

Falcon, L. M., & Melendez, E. (1996, February). *The role of social networks in the labor market outcomes of Latinos, blacks and non-Hispanic whites.* Boston: University of Massachusetts-Boston, Boston Urban Inequality Research Group.

Ferguson, R. F. (1996). Shifting challenges: Fifty years of economic change towards black-white earnings equality. In O. Clayton, Jr. (Ed.), *An American dilemma revisited: Race relations in a changing world.* New York: Russell Sage.

Fitzgerald, J. (1995). *Making school-to-work happen in inner cities: A white paper prepared for the MacArthur Foundation.* Chicago: University of Illinois at Chicago, Great Cities Institute.

Fitzgerald, J., & Jenkins, D. (1997, January). *Best-practice in community colleges: Connecting the urban poor to education and employment opportunities.* Chicago: University of Illinois at Chicago, Great Cities Institute.

Freeman, R. B., & Katz, L. B. (1994). Rising wage inequality: The United States vs. other advanced countries. In R. B. Freeman (Ed.), *Working under different rules.* New York: Russell Sage Foundation.

Ganz, M. (1995, November). *Strategic innovation, social movements and institutional change: Unionization in California agriculture (1947-1977).* Unpublished manuscript, Harvard University, Department of Sociology, Cambridge, MA.

Gerlach, M. L. (1992). *Alliance capitalism: The social organization of Japanese business.* Berkeley: University of California Press.

Giddens, A. (1984). *The construction of society: Outline of the theory of structuration.* Berkeley: University of California Press.

Gittell, M., Gross, J., & Newman, K. (1994). *Race and gender in neighborhood development organizations.* New York: City University of New York, Howard Samuels State Management and Policy Center, Graduate Center.

Gittell, R., & Wilder, M. (1995, May). *Best practices in community revitalization* (p. 16). Durham, University of New Hampshire, Whittemore School of Business.

Glickman, N. J., & Nye, N. (1996, January 23). *Understanding the critical roles of community development partnerships and collaboratives in the community de-*

velopment process. New Brunswick, NJ: Rutgers University, Center for Urban Policy Research.

Gordon, D. M. (1971). *Theories of poverty and underemployment*. Lexington, MA: D. C. Heath.

Gordon, D. M., Edwards, R., & Reich, M. (1982). *Segmented work, divided workers*. New York: Oxford University Press.

Gottschalk, P., & Moffitt, R. (1994). The growth of earnings instability in the U.S. labor market. *Brookings Papers on Economic Activity, 2*, 217-272.

Granovetter, M. (1985, November). Economic action and social structure: The problem of embeddedness. *American Journal of Sociology*, 481-510.

Granovetter, M. (1994). The sociological and economic approaches to labor market analysis. In G. Farkas, & R. Egland (Eds.), *Industries, firms, and jobs* (expanded edition). Hawthorn, NY: Aldine.

Granovetter, M., & Tilly, C. (1988). Inequality and labor processes. In N. J. Smelser (Ed.), *Handbook of sociology*. Newbury Park, CA: Sage.

Harrison, B. (1972). *Education, training and the urban ghetto*. Baltimore, MD: Johns Hopkins University Press.

Harrison, B. (1994). *Lean and mean: The changing landscape of corporate power in the age of flexibility*. New York: Basic Books.

Harrison, B., Weiss, M., & Gant, J. (1995). *Building bridges: CDCs and the world of employment training*. New York: Ford Foundation.

Henckoff, R. (1993, March 22). Companies that train best. *Fortune*, p. 62.

Hershey, A., & Rosenberg, L. (1994, June). *The study of the replication of the CET job training model*. Washington, DC: Mathematica.

Hershey, R. D., Jr. (1995, November 1). Working earnings post rise of 2.7%, lowest on record. *New York Times*, p. A1.

Higdon, F. X. (1993). *Job links program evaluation*. Unpublished manuscript, University of Pittsburgh, Graduate School of Public and International Affairs.

Hill, E. W. (1996, February). *Revitalizing Cleveland, not comeback Cleveland: Local and federal forces that rebuild a region*. Cleveland: Cleveland State University, Maxine Goodman Levin College of Urban Affairs.

Hill, I. (1995). *Made in Brooklyn* [Video]. New Day Films.

Hollister, R. G. (1990). *The minority female single parent demonstration: New evidence about effective training strategies*. New York: Rockefeller Foundation.

Hollister, R. G., & Hill, J. (1995, April). *Problems in the evaluation of community-wide initiatives* (Working Paper No. 70). New York: Russell Sage Foundation.

Holzer, H. J. (1987, June). Informal job search and black youth unemployment. *American Economic Review*, 446-452.

Holzer, H. J. (1996). *What employers want: Job prospects for less-educated workers*. New York: Russell Sage Foundation.

Howell, D. (1994, Summer). The skills myth. *The American Prospect*, 81-89.

Howell, D. (1997, February). *The collapse of low-skill wages: Technological shift or institutional failure?* [Public policy brief]. Annandale-on-Hudson, NY: Bard College, The Jerome Levy Economics Institute.

J. A. Croson Company v. City of Richmond, 488 U.S. 469 (1989).

Jacoby, S. M. (1985). *Employing bureaucracy: Managers, unions, and the transformation of work in American industry, 1900-1945*. New York: Columbia University Press.

Jobs for the Future. (1995). *School-to-work and community economic development: Identifying common ground.* Boston: Author.

Kanter, R. M. (1995). *World class: Thriving locally in the global economy.* New York: Simon & Schuster.

Kasarda, J. D. (1995). Industrial restructuring and the changing location of jobs. In R. Farley (Ed.), *State of the union: America in the 1990s* (Vols. I and II). New York: Russell Sage Foundation.

Keating, W. D., Krumholz, N., & Metzger, J. (1995). Postpopulist public-private partnerships. In W. D. Keating, N. Krumholz, & D. C. Perry (Eds.), *Cleveland: A metropolitan reader.* Kent, OH: Kent State University Press.

Keler, K., & Lange, J. (1995, June). Community group joins with union to fight inner-city poverty in Baltimore. *Labor Notes,* p. 3.

Kelley, M. R., & Arora, A. (1996, March). The role of institution-building in U.S. industrial modernization programs. *Research Policy, 25*(2), 265-280.

Kennedy, R. (1994, August 28). Connections, connections: Across the city, strategist guide small businesses toward the economic development frontier. *Sunday New York Times,* p. A1.

Kerachsky, S. (1994). *The minority female single parent demonstration: Making a difference—Does an integrated program model promote more jobs and higher pay?* Washington, DC: Mathematica Policy Research, Inc.

Kirschenman, J., & Neckerman, K. M. (1991). "We'd like to hire them, but . . ." The meaning of race to employers. In C. Jencks & P. E. Peterson (Eds.), *The urban underclass.* Washington, DC: Brookings Institution.

Knocke, D., & Kalleberg, A. L. (1994, August). Job training in U.S. organizations. *American Sociological Review,* 537-546.

Kotlowitz, A. (1991, September 17). Community groups quietly make strides in inner-city housing. *Wall Street Journal,* p. A1.

Krugman, P. (1992, Fall). The right, the rich, and the facts. *The American Prospect,* 19-31.

Levy, F. (1995). Income and income inequality since 1970. In R. Farley (Ed.), *State of the union: America in the 1990s, Volume I: Economic trends.* New York: Russell Sage Foundation.

Levy, F., & Murnane, R. J. (1992, Summer). Where will all the smart kids work? *Journal of the American Planning Association,* 283-287.

Lin, N. (1982). Social resources and instrumental action. In P. V. Marsden & N. Lin (Eds.), *Social structure and network analysis.* Beverly Hills, CA: Sage.

Lin, N., Ensel, W. M., & Vaughn, J. C. (1981, August). Social resources and strength of ties: Structural factors in occupational status attainment. *American Sociological Review.*

Lunt, P. (1993, August). Urban allies on the move. *American Bankers Association Banking Journal,* 34-38.

Lynch, L. M. (1993). Payoffs to alternative training strategies at work. In R. B. Freeman (Ed.), *Working under different rules.* New York: Russell Sage Foundation.

Mangum, G. (1995). Nostra culpae: A critique of 33 years of employment, training, and welfare programs. In M. Pines, G. Mangum, & B. Spring (Eds.), *The harassed staffer's guide to employment and training policy.* Baltimore, MD: Johns Hopkins University, Sar Levitan Center for Social Policy Studies.

Mangum, S. (1995, June). The employment outlook for troubled populations. In M. Pines, G. Mangum, & B. Spring (Eds.), *The harassed staffers' guide to employment and training policy*. Baltimore, MD: Johns Hopkins University, Sar Levitan Center for Social Policy Studies.

Marcotte, D. (1994). *Evidence of a fall in the wage premium and job security*. Dekalb: Northern Illinois University, Center for Governmental Studies.

Marshall, R., & Tucker, M. (1992). *Thinking for a living: Education and the wealth of nations*. New York: Basic Books.

McCormick, L. (1996, May). *Strategic manufacturing network alliances in Chicago's metalworking industries in the 20th century*. Unpublished doctoral dissertation, Massachusetts Institute of Technology, Department of Urban Studies and Planning, Cambridge, MA.

Medoff, P., & Sklar, H. (1994). *Streets of hope: The fall and rise of an urban neighborhood* (pp. 75-77). Boston: South End.

Melendez, E. (1996). *Working on jobs: The Center for Employment Training*. Boston: University of Massachusetts-Boston, Gaston Institute.

Metzger, J. T. (1992). The CRA and neighborhood revitalization in Pittsburgh. In G. D. Squires (Ed.), *From redlining to reinvestment*. Philadelphia: Temple University Press.

Mincy, R. B., & Weiner, S. J. (1993, July). The underclass in the 1980s: Changing concept, constant reality (Working paper). Washington, DC: Urban Institute.

Mishel, L. (1995, Fall). Rising tides, sinking wages. *The American Prospect*, 60-64.

Mishel, L., & Bernstein, J. (1994). *The state of working America, 1994-95*. Armonk, NY: M. E. Sharpe (for the Economic Policy Institute).

Mishel, L., & Teixera, R. A. (1991, Fall). The myth of the coming labor shortage. *The American Prospect*, 98-103.

Montgomery, J. D. (1991, December). Social networks and labor-market outcomes: Towards an economic analysis. *American Economic Review*, 1408-1418.

Montgomery, J. D. (1992, October). Job search and network composition: Implications of the strength-of-weak-ties hypothesis. *American Sociological Review*, 586-596.

Morales, R., & Bonilla, F. (Eds.). (1993). *Latinos in a changing U.S. economy*. Newbury Park, CA: Sage.

Mortenson, D. T. (1986). Job search and labor market analysis. In O. Ashenfelter & R. Layard (Eds.), *Handbook of labor economics*. Amsterdam: North-Holland.

Moss, P., & Tilly, C. (1996a, August). "Soft" skills and race: An investigation of black men's employment problems. *Work and Occupations*, 23(3), 252-276.

Moss, P., & Tilly, C. (1996b, November). *Informal hiring practices, racial exclusion, and public policy*. Unpublished manuscript, University of Massachusetts at Lowell, Department of Policy and Planning.

Nye, N., & Glickman, N. J. (1995, November). *Expanding local capacity through community development partnerships*. New Brunswick, NJ: Rutgers University, Center for Urban Policy Research.

Oliver, M. L. (1988, October). The urban black community as network: Towards a social network perspective. *Sociological Quarterly*, 9(4), 623-645.

Osterman, P. (Ed.). (1984). *Internal labor markets*. New York: Oxford University Press.

Osterman, P. (1988). *Employment futures*. New York: Oxford University Press.

Osterman, P. (1994, January). How common is workplace transformation and how can we explain who does it? *Industrial and Labor Relations Review*, pp. 173-188.

Osterman, P. (1995, April). Skills, training, and work organization in American establishments. *Industrial Relations*, pp. 125-146.

Osterman, P., & Batt, R. (1993, Summer). Employer centered training programs for international competitiveness: Lessons from state programs. *Journal of Policy Analysis and Management*, 456-477.

Osterman, P., & Lautsch, B. A. (1996, January). *PROJECT QUEST: A report to the Ford Foundation*. Cambridge: Massachusetts Institute of Technology, Sloan School of Management.

Owens, M. L. (1996, March). *Race, place, and government employment*. Albany: State University of New York at Albany, Nelson A. Rockefeller Institute of Government.

Peck, J. (1996). *Work place: The social regulation of labor markets*. New York: Guilford.

Peschek, B. (1997, March/April). A living wage? Campaigns attach strings to public contracts. *Dollars and Sense*, pp. 28-34.

Pfeffer, J., & Baron, J. N. (1988). Taking the workers back out: Recent trends in the structuring of employment. In B. Staw & L. L. Cummings (Eds.), *Research in organizational behavior* (Vol. 10). Greenwich, CT: JAI.

Pierce, N. R., & Steinbach, C. F. (1987). *Corrective capitalism: The rise of America's CDCs*. New York: Ford Foundation.

Pines, M., Mangum, G., & Spring, B. (Eds.). (1995, August). *The harassed staffer's guide to employment and training policy*. Baltimore, MD: Johns Hopkins University, Sar Levitan Center for Social Policy Studies.

Piore, M. J. (1995). *Beyond individualism*. Cambridge, MA: Harvard University Press.

Piore, M. J., & Sabel, C. F. (1984). *The second industrial divide*. New York: Basic Books.

Pouncy, H., & Mincy, R. B. (1995). Out of welfare: Strategies for welfare-bound youth. In D. Smith Nightingale & R. H. Haverman (Eds.), *The work alternative: Welfare reform and the realities of the job market*. Washington, DC: Urban Institute Press.

Powell, W. W., & Smith-Doerr, L. (1994). Networks and economic life. In N. J. Smelser & R. Swedberg (Eds.), *The handbook of economic sociology*. Princeton, NJ/New York: Princeton University Press/Russell Sage Foundation.

Reardon, C. (1995, Summer/Fall). A living wage: San Antonio's churches breathe life into a troubled economy. *Ford Foundation Report*, 7.

Rees, A. J., Jr. (1966, May). Information networks in labor markets. *American Economic Review/Proceedings*, 559-566.

Rees, A. J., Jr., & Schultz, G. P. (1970). *Workers in an urban labor market*. Chicago: University of Chicago Press.

Rogard-Tabori, J. (1995). *Project design and evaluation guidebook*. Washington, DC: BHM International.

Rogers, M. B. (1990). *Cold anger: A story of faith and power politics*. Denton: University of North Texas Press.

Romer, P. (1986). Increasing returns and long-run growth. *Journal of Political Economy, 94*, 1002-1037.

Rosalso, R. (1989). *Culture and truth: The remaking of social analysis.* Boston: Beacon.

Rose, S. (1995). *The decline of employment stability in the 1980s.* Washington, DC: National Commission on Employment Policy.

Rosenfeld, S. (1994). *Two-year colleges at the forefront: The consortium for manufacturing competitiveness.* Chapel Hill, NC: Regional Technology Strategies.

Rosenfeld, S. A. (1995a). *Industrial-strength strategies: Regional business clusters and public policy.* Washington, DC: The Aspen Institute, Rural Economic Policy Program.

Rosenfeld, S. A. (Ed.). (1995b). *New technologies and new skills: Two-year colleges at the vanguard of modernization.* Chapel Hill, NC: Regional Technology Strategies.

Rosenfeld, S. A., & Kingslow, M. E. (1995). *Advancing opportunity in advanced manufacturing: The potential of predominantly two-year colleges.* Chapel Hill, NC: Regional Technology Strategies.

Rosenthal, N. H. (1995, June). The nature of occupational employment growth: 1983-93. *Monthly Labor Review,* pp. 45-54.

Rusk, D. (1993). *Cities without suburbs.* Washington, DC: Woodrow Wilson Center Press.

Sabel, C. F. (1996, March). A measure of federalism: Assessing manufacturing technology centers. *Research Policy, 25*(2), 281-308.

Sassen, S. (1989, Fall). America's immigration problem. *World Policy, 6,* 811-832.

Savner, S. (1996, January 15). *Devolution, workforce development, and welfare reform.* Washington, DC: Center for Law and Social Policy.

Saxenian, A-L. (1994). *Regional advantage: Silicon Valley and Route 128.* Cambridge, MA: Harvard University Press.

Taub, R. P. (1988). *Community capitalism.* Cambridge, MA: Harvard University Press.

Taub, R. P., Surgeon, G. P., Lindholm, S., Betts Otti, P., & Bridges, A. (1977, September). Urban voluntary associations, locality based and externally induced. *American Journal of Sociology, 83,* 425-442.

Thompson, J. P. (1997, February). *Universalism and deconcentration: Why race still matters in poverty and economic development.* New York: Barnard College, Department of Political Science.

Tilly, C. (1995). *Half a job.* Philadelphia: Temple University Press.

Tilly, C. (1996, June). *The good, the bad, and the ugly: Good and bad jobs in the United States at the millennium.* New York: Russell Sage Foundation.

Tilly, C., & Tilly, C. (1994). Capitalist work and labor markets. In N. J. Smelser & R. Swedberg (Eds.), *The handbook of economic sociology.* Princeton, NJ/New York: Princeton University Press/Russell Sage Foundation.

U.S. Department of Labor, Bureau of Labor Statistics. (n.d.). *Occupational outlook handbook.* Washington, DC: Author. (Published every 2 years)

U.S. Department of Labor, Office of the Chief Economist. (1995, January). *What's working (and what's not): A summary of research on the economic impacts of employment and training programs.* Washington, DC: U.S. Government Printing Office.

Uzzi, B. (in press). Embeddedness and economic performance: The network effect. *American Sociological Review.*

Vietorisz, T., & Harrison, B. (1970). *The economic development of Harlem.* New York: Praeger.

Waldinger, R. (1996). *Still the promised city?* Cambridge, MA: Harvard University Press.

Waquant, L. J. D. (1994). The new urban color line: The state and fate of the ghetto in post-Fordist America. In C. Calhoun (Ed.), *Social theory and the politics of identity* (pp. 231-276). Cambridge, MA: Blackwell.

Webster's new collegiate dictionary. (1979). Springfield, MA: Merriam.

Weiss, M. A., & Metzger, J. T. (1987, Autumn). Technological development, neighborhood planning, and negotiated partnerships: The case of Pittsburgh's Oakland neighborhood. *Journal of the American Planning Association, 53*(4), 469-477.

Wolff, E. N. (1995, Summer). How the pie is sliced: America's growing concentration of wealth. *American Prospect,* pp. 58-64.

Wylde, K. (1996, Winter). Doing the right thing. *City Journal,* 6(1).

Zabrowski, A., & Gordon, A. (1993). *Evaluation of minority female single parent demonstration: Fifth year impacts at CET.* New York: Mathematica Policy Research (for the Rockefeller Foundation).

Zemsky, R. (1994). *What employers want: Employer perspectives on youth, the youth labor market, and prospects for a national system of youth apprencticships* (Working Paper No. WP22). Philadelphia: University of Pennsylvania, National Center on the Educational Quality of the Workforce.

INDEX

ABOUT THE AUTHORS

Bennett Harrison is Professor of Urban Political Economy in the Milano Graduate School of Management and Urban Policy at the New School for Social Research in New York City and an associate of the school's Community Development Research Center. He teaches urban and regional economic development, labor economics, and poverty and discrimination. Formerly, he taught at Massachusetts Institute of Technology, Carnegie Mellon University, and Harvard's Kennedy School of Government. He is author or coauthor of 10 books and approximately 100 papers in scholarly journals and is a frequent contributor of opinion editorials to the *New York Times, U.S. News & World Report,* and other media. His recent books are *Building Bridges: Community Development Corporations and the World of Employment Training* (with Marcus Weiss and Jon Gant) and *Lean and Mean: The Changing Landscape of Corporate Power in the Age of Flexibility.* He holds a doctorate in economics from the University of Pennsylvania and a bachelor's degree from Brandeis University. He is especially well-known, nationally and internationally, for his two books on industrial restructuring and the polarization of income coauthored with Barry Bluestone: *The Deindustrialization of America* and *The Great U-Turn.* He is currently conducting research on networks, metropolitan economies, and jobs.

Marcus Weiss is President of the Economic Development Assistance Consortium, a consulting firm in Boston, Massachusetts. He specializes in commercial revitalization, innovative linkages between training and economic development, community reinvestment, community development finance institutions, minority business initiatives and policy and legislative issues of concern to government, community-based organizations, and foundations. He is coauthor of *Building Bridges: Community Development Corporations and the World of Employment Training* (Ford Foundation) and of *Community Reinvestment Act: How to Implement Your Bank's Program* (Sheshunoff Information Services) and editor of *Community Reinvestment Act: Access to Development Capital/Operating in a Changing Regulatory Environment* (Massachusetts Continuing Legal Education). He also serves as legal counsel to the Massachusetts Association of Community Development Corporations and as a regular consultant to the National League of Cities, the U.S. Department of Housing and Urban Development, and the U.S. Department of Commerce. He holds a law degree from Boston University School of Law and a bachelor's degree in political science from the American University School of Government.